Analysis and Separations of Trace Elements in the Environment

Analysis and Separations of Trace Elements in the Environment

Topic Editors

Gene Hall
Begoña González

MDPI

Basel • Beijing • Wuhan • Barcelona • Belgrade • Novi Sad • Cluj • Manchester

Topic Editors

Gene Hall
Department of Chemistry and Chemical Biology
The State University of New Jersey
New Brunswick
USA

Begoña González
Department of Chemical Engineering
University of Vigo
Vigo
Spain

Editorial Office
MDPI AG
Grosspeteranlage 5
4052 Basel, Switzerland

This is a reprint of the Topic, published open access by the journals *Minerals* (ISSN 2075-163X), *Molecules* (ISSN 1420-3049), *Separations* (ISSN 2297-8739) and *Water* (ISSN 2073-4441), freely accessible at: https://www.mdpi.com/topics/D12O35T268.

For citation purposes, cite each article independently as indicated on the article page online and as indicated below:

Lastname, A.A.; Lastname, B.B. Article Title. *Journal Name* **Year**, *Volume Number*, Page Range.

ISBN 978-3-7258-7601-3 (Hbk)
ISBN 978-3-7258-7602-0 (PDF)
https://doi.org/10.3390/books978-3-7258-7602-0

Contents

About the Editors

Gene Hall

Gene Hall is a senior professor of analytical chemistry in the Department of Chemistry and Chemical Biology at The State University of New Jersey, where he has over 44 years of teaching and research experience. He received his Ph.D. in radiochemistry from Virginia Tech in 1978 and has long been dedicated to research in applied analytical chemistry, particularly renowned for his work in product forensics. He utilizes advanced mass spectrometry and chromatography techniques (such as ICP-MS, Q-TOF, and UHPLC) for molecular fingerprinting of dietary supplements (such as Omega-3 fish oil and CBD), counterfeit consumer products, and e-cigarette ingredients. He is also known for his research on heavy metal (such as lead) tracing in the environmental field and his contributions to public health. Within the university, he is renowned for teaching challenging instrumental analysis courses and is a highly regarded and rigorous engineering educator.

Begoña González

Begoña González works at the FEQx Lab, Department of Chemical Engineering, University of Vigo. He received a nomination for a position at the university in April 2024. His research areas include textile waste treatment, polymer material recycling and separation technology, thermal analysis and calorimetry of polyphenols, and the application of green solvents. In teaching, he is responsible for supervising undergraduate graduation projects at the university. His research directions include the extraction of antioxidants using green solvents, the extraction of phenolic compounds using biosolvents, and the removal of colored waste using DES. Academically, he collaborates extensively with scholars such as Pablo B. Sánchez and Lorena Villar at the university and has published numerous research papers in international core journals such as *Molecules* and *Fibers and Polymers*. His core achievements include the determination of the thermophysical properties of polyphenols, the development of hydrolysis-assisted fractionation technology for cotton/polyester blended textile waste, and the treatment of wastewater dyes using environmentally friendly microspheres.

molecules

MDPI

Article

Efficient Analytical Pretreatment of Cr(VI) in Ethylene Wastewater by Grafting g-C_3N_4 Material Based on Coupling Agent-Modified Basalt Matrix (Basalt–MTES/g-C_3N_4)

Zheng Wang, Mingchang Jia *, Yi Ren, Hongmin Ren, Shuhao Liang, Jiaru Sun, Siqi Hao, Jinchuan Li and He Li

Department of Chemical Engineering, Hebei Petroleum University of Technology, Chengde 067000, China; 18789286658@163.com (Z.W.); 15532769658@163.com (Y.R.); 15831463175@163.com (H.R.); 15032363962@163.com (S.L.); 17320912605@163.com (J.S.); haosqcdpc123@yeah.net (S.H.); 17733750705@163.com (J.L.); 18832423881@163.com (H.L.)

* Correspondence: 15512959617@126.com

Abstract: This study presents a novel Basalt-based grafted graphitic carbon nitride composite (Basalt–MTES/g-C_3N_4) for the efficient pretreatment of Cr(VI) in ethylene wastewater. The composite was synthesized by the acid purification of natural Basalt, surface modification with hydroxymethyl triethoxysilane (MTES), and the subsequent grafting of g-C_3N_4. Characterization confirmed the uniform distribution of nano-sized g-C_3N_4 particles on a Basalt surface with intact chemical bonding, where 82.63% of melamine participated in g-C_3N_4 crystallization. The material exhibited a high specific surface area (403.55 m^2/g) and mesoporous structure (34.29 nm). Acidic conditions promoted the protonation of amino groups in g-C_3N_4, significantly enhancing Cr(VI) adsorption via ion exchange. Adsorption kinetics followed the pseudo-second-order model, while isotherm data fitted the Langmuir monolayer adsorption mechanism. The composite achieved 97% Cr(VI) recovery through chromatographic extraction and retained 96.87% removal efficiency after five regeneration cycles. This work demonstrates a cost-effective, recyclable green pretreatment material for high-sensitivity Cr(VI) monitoring in ethylene industry wastewater, offering dual benefits in environmental remediation and regulatory compliance. The design synergizes natural Basalt's stability with g-C_3N_4's adsorption affinity, showing practical potential for sustainable wastewater treatment technologies.

Keywords: basalt–MTES/g-C_3N_4; Cr(VI); ethylene wastewater; analytical pretreatment; ion exchange

1. Introduction

As the core product of the modern petrochemical industry, ethylene's production scale directly reflects the industrialization level of a country. The global ethylene production capacity has been continuously increasing in recent years, reaching 210 million tons in 2023, and is widely used in plastics, synthetic rubber, fibers, and other fields [1]. However, at the same time, the environmental monitoring issues involved in the ethylene refining process have to be given increasing attention. According to statistics, as much as 18.2 billion tons of wastewater is generated globally each year from the refining of ethylene (2017) [2]. The industrial wastewater produced during the ethylene refining process is complex in composition, containing a large amount of refractory organic matter and heavy metal ions, posing a serious threat to the ecosystem and human health [3]. Although various countries have set strict standards for the discharge of ethylene wastewater (for instance, China's 'Pollutant Discharge Standard for Petrochemical Industry' stipulates that the COD limit is 60 mg·L^{-1}), there are still problems, such as the strong concealment of pollutants and insufficient detection sensitivity in actual monitoring [4]. In particular, the trace analysis of highly toxic substances like Cr(VI) has become difficult in environmental monitoring [5].

In the monitoring of heavy metals in ethylene wastewater, Cr(VI) is listed as a priority pollutant to be controlled due to its strong oxidizing property, carcinogenicity, and bioaccumulation [6]. Current standards (such as the EU's "Industrial Emission Directive") require that the emission concentration of Cr(VI) shall not exceed 0.01 mg·L^{-1}, which is much lower than the limits of metals such as iron (2.0 mg·L^{-1}) and zinc (1.0 mg·L^{-1}) [7]. However, Cr(VI) in ethylene wastewater often coexists with Fe^{3+}, Cu^{2+}, etc., and is disturbed by factors such as pH and organic matter [8]. Traditional spectrophotometric methods (such as the diphenylcarbazide method) are prone to false positivity or insufficient sensitivity [9]. Therefore, developing efficient enrichment and selective separation techniques has become the key to achieving the precise analysis of trace Cr(VI).

At present, the analytical pretreatment of Cr(VI) in ethylene production enterprises (such as the Sinopec Yanshan Petrochemical Company) mainly relies on the evaporation concentration method (simple principle but cumbersome operation), the ion exchange method (difficult resin regeneration), and the membrane enrichment method (severe membrane fouling) [10,11]. Although the extraction chromatography method has excellent selectivity, it requires the synthesis of specific functionalized eluent resins (such as those containing amino chelating groups), which makes the process complex and costly [12]. In recent years, graphite carbon nitride (g-C_3N_4) has shown a specific treatment capacity for Cr(VI) due to its layered structure rich in Lewis bases. Most applications of g-C_3N_4 in Cr(VI) treatment mainly focus on the reduction in Cr(VI) under photocatalysis, and studies have shown that based on the possible ion exchange or coordination complexation between the g-C_3N_4 functional group and $Cr_2O_7{}^{2-}$ or $CrO_4{}^{2-}$, it also has great potential in the adsorption and analytical pretreatment applications of Cr(VI) [13,14]. However, its powder form is prone to agglomeration and difficult to recover, which limits its practical application. If it is immobilized on the surface of the porous matrix material, it can not only retain the adsorption activity but also improve the stability of the material. This idea provides new thoughts for the design and synthesis of new analytical pretreatment materials.

In recent studies, Basalt has become an ideal matrix for functional materials due to its natural porous structure, high mechanical strength, and surface-active groups [15]. Firstly, its natural porous structure (with pore diameters ranging from 20 to 80 nm) significantly enhances the adsorption capacity and rate, which is because it provides a high specific surface area (up to over 550 m^2·g^{-1}) and abundant mass transfer channels [16]. It is not uncommon for the matrix of porous structural materials to provide excellent reactive performance. For example, in the study of mesoporous carbon CMK-3 encapsulating nickel nanoparticles and homopoly (vinylsulfonic acid) as acid–metal bifunctional reduction amination catalysts by Kalbasi R. et al., the excellent pore structure played a significant promoting role in the catalytic effect [17]. Secondly, Basalt has excellent mechanical strength (tensile strength $\geq$ 3000 MPa) and thermal stability (temperature resistance > 600 °C) [18]. It can maintain structural integrity even under dynamic water flow or complex working conditions, avoiding secondary pollution caused by material breakage. The chemical inertness characteristic enables it to remain stable in the industrial wastewater environment, thereby extending the service life of the material. Furthermore, the abundant hydroxyl groups (with a -OH density of approximately 3.2/nm^2) on the surface of Basalt can be directionally modified by silane coupling agents, providing an ideal interface for the loading of functional materials (such as g-C_3N_4) and achieving the precise regulation of adsorption performance [19]. In this study, Basalt material was used as the framework. The Basalt material was modified by the coupling agent hydroxymethyl triethoxysilane (MTES), and g-C_3N_4 was successfully grafted to construct a composite material (Basalt–MTES/g-C_3N_4) with both adsorption and separation functions for the pretreatment process of Cr(VI) analysis and measurement in ethylene wastewater. The surface conformation and synthesis mechanism of Basalt–MTES/g-C_3N_4 were analyzed by characterization methods such as SEM, FTIR, XRD, XPS, and BET. In terms of the application of synthetic materials, this study systematically evaluated the effects of Basalt–MTES/g-C_3N_4 on the acidity, isotherms, kinetics, adsorption heat, and regeneration performance of the Cr(VI) adsorption process, applied it to the analytical pretreatment of actual ethylene wastewater, and significantly improved the detection accuracy of the flame atomic absorption spectrophotometer. This research provides a design idea of new pretreatment materials for the high-sensitivity analysis of Cr(VI) in ethylene wastewater, which is of great practical significance for promoting the green production and environmental supervision technology progress in the ethylene industry.

The synthesis flowchart of Basalt–MTES/g-C_3N_4 is shown in Figure 1.

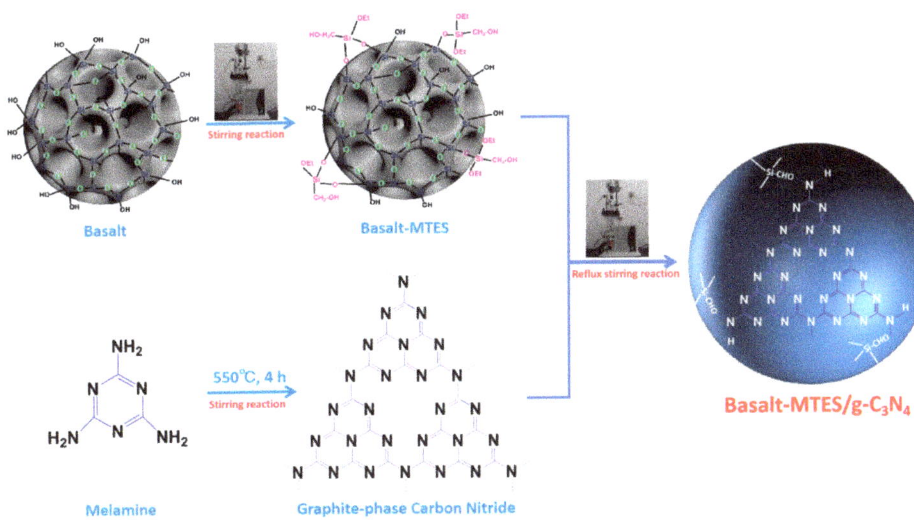

Figure 1. Synthesis process of Basalt–MTES/g-C_3N_4.

2. Results and Discussion

2.1. Characterization of Basalt, Basalt–MTES, and Basalt–MTES/g-C_3N_4

The SEM characterization of Basalt, Basalt–MTES, and Basalt–MTES/g-C_3N_4 is shown in Figure 2. For the Basalt (Figure 2a), since it has been thoroughly treated with HNO_3, the soluble impurity components in the Basalt material are removed. Through this process, a large number of pores were formed on the surface and inside the material, and the specific surface area was greatly increased compared to before. After removing the impurities, the surface of the Basalt showed a lamellar porous structure. For Basalt–MTES (Figure 2b), because the grafting reaction of the coupling agent belongs to the small-molecule grafting reaction, the appearance of Basalt did not change significantly after the introduction of the coupling agent MTES. However, after further grafting g-C_3N_4, the surface of the material changed significantly. Since the formation process of carbon nitride belongs to the polymerization growth reaction, g-C_3N_4 will eventually be grafted on the surface of the matrix material in the form of a polymer, and the surface of the material will be covered. As shown in Figure 2c, the nano-sized granular material on the surface is a g-C_3N_4 polymer particle [20]. Furthermore, based on visual inspection, the diameter of the g-C_3N_4 particles is mostly less than 100 nm.

Figure 2. SEM images of Basalt (**a**), Basalt–MTES (**b**), and Basalt–MTES/g-C_3N_4 (**c**).

The FTIR spectra of the Basalt, Basalt–MTES, and Basalt–MTES/g-C_3N_4 are shown in Figure 3. For the Basalt (Figure 3a), 1062 cm^{-1} is the stretching vibration peak of Si-O-Si, which is because Basalt is mainly composed of SiO_2 (45–52%) and a variety of metal oxides (Al_2O_3, FeO, Fe_2O_3, CaO, MgO, etc.). After advanced treatment with HNO_3, the metal oxides were dissolved, and the main framework of SiO_2 was retained [21]. Overall, 971 cm^{-1} is the flexural vibration peak of -OH, 1627 cm^{-1} is the in-plane deformation vibration peak of -OH, and the broad peak around 3430 cm^{-1} is the stretching vibration peak of the double-base -OH and H_2O [22,23]. This information indicates that the acid treatment process retained the SiO_2 framework of Basalt and shows that after the acidification of HNO_3, the surface of Basalt is rich in -OH.

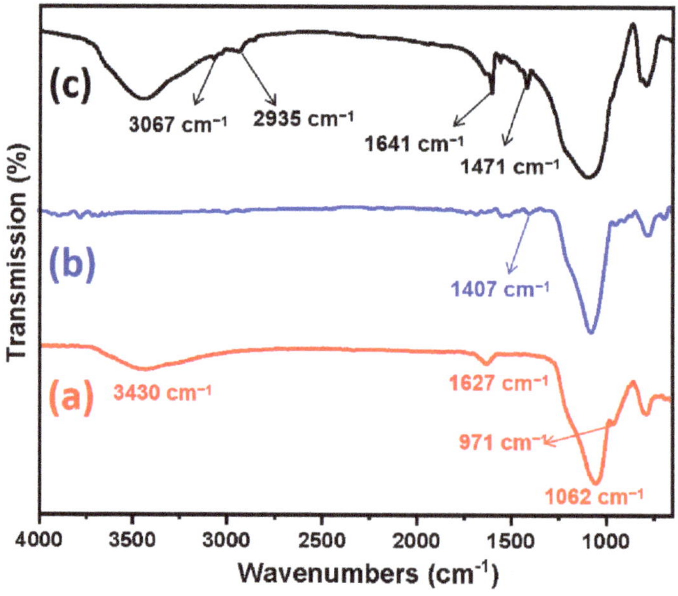

Figure 3. FTIR spectra of Basalt (**a**), Basalt–MTES (**b**), and Basalt–MTES/g-C_3N_4 (**c**).

For the Basalt–MTES (Figure 3b), the flexural vibration peak of -OH at 971 cm^{-1} and the in-plane deformation vibration peak of -OH at 1627 cm^{-1} were significantly weakened or disappeared, which indicates that the large amount of -OH on the surface of Basalt may react with the coupling agent MTES, and in this reaction, the -OCH_3 in the coupling agent molecule condenses with the -OH on the surface of Basalt, and the two are chemically bonded together in the form of Si-O-Si [24]. The reaction is as follows:

The wide peak at 3430 cm^{-1} disappears, indicating the disappearance of moisture on the surface of Basalt; that is, there may be a large number of organic hydrophobic groups (coupling agent molecules) on the surface of Basalt [25]. The weak peak at 1407 cm^{-1} is the in-plane bending vibration peak of C-H in -CH_3 (due to the single-layer grafting of the coupling agent molecules, the peak intensity is relatively weak). The above information initially indicates that the coupling agent MTES has been grafted onto the surface of Basalt.

For the Basalt–MTES/g-C_3N_4 (Figure 3c), the characteristic peak at 1471 cm^{-1} is the C-C stretching vibration in the melamine ring outside, and the characteristic peak of 1641 cm^{-1} is the C-N stretching vibration in the melamine inside the ring [26]. In addition, 3067 cm^{-1} is the stretching vibration peak of C-H on -CH_2-, and 2935 cm^{-1} is the stretching vibration peak of C-H in -CH_2OH. This peak position information indicates that melamine has successfully grafted and polymerized on the material surface.

The XRD analysis results of the Basalt, Basalt–MTES, and Basalt–MTES/g-C_3N_4 are shown in Figure 4. For Basalt (Figure 4a), there is only one main characteristic peak of 22.36° in the spectrum, which is a typical feature peak of amorphous SiO_2. This indicates that only SiO_2 remains in the Basalt matrix substrate after acid treatment [27]. After grafting the coupling agent MTES, for the spectra of Basalt–MTES (Figure 4b), there are clearly organic compound diffraction peaks at 30.72° and 31.96°, which indicates that the coupling agent molecules are already present on the surface of the material [28]. For Basalt–MTES/g-C_3N_4 (Figure 4c), characteristic peaks appeared at 20.16° and 27.44°. The diffraction peak at 20.1° belongs to the microcrystalline structure of g-C_3N_4, and the diffraction peak at 27.4° corresponds to the (002) crystal plane, which belongs to the interlayer superposition of g-C_3N_4 [29]. This indicates that g-C_3N_4 and Basalt–MTES have successfully recombined. This analysis result is basically consistent with the FTIR result.

The XPS spectrogram of Basalt, Basalt–MTES, and Basalt–MTES/g-C_3N_4 is shown in Figure 5a, and the N 1s spectrum of Basalt–MTES/g-C_3N_4 is shown in Figure 5b. In Figure 5a, for Basalt (a_1), there are only two elements: Si and O. Among them, the binding energy 103 eV is the Si 2p absorption peak, 153 eV is the Si 2s absorption peak, and 533 eV is the O 1s absorption peak, which indicates that the main framework component of the Basalt matrix after acid treatment is SiO_2 [30]. Compared with Basalt, for the Basalt–MTES (a_2) spectrogram, in addition to the Si and O absorption peaks, there is also the C 1s absorption peak, at 285 eV, indicating that the coupling agent molecules may already exist on the surface of Basalt [31]. For Basalt–MTES/g-C_3N_4 (a_3), besides the above-mentioned peaks, the absorption peak of N 1s appears at 399 eV, which indicates that the melamine was successfully grafted to the surface of the Basalt [32].

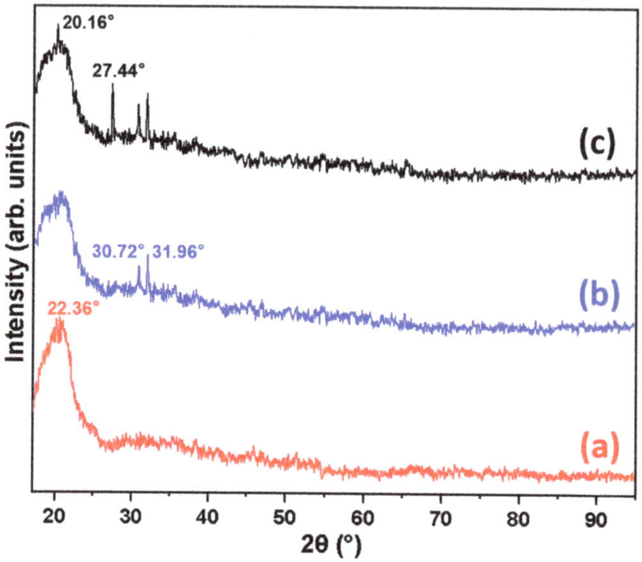

Figure 4. XRD pattern of Basalt (**a**), Basalt–MTES (**b**), and Basalt–MTES/g-C_3N_4 (**c**).

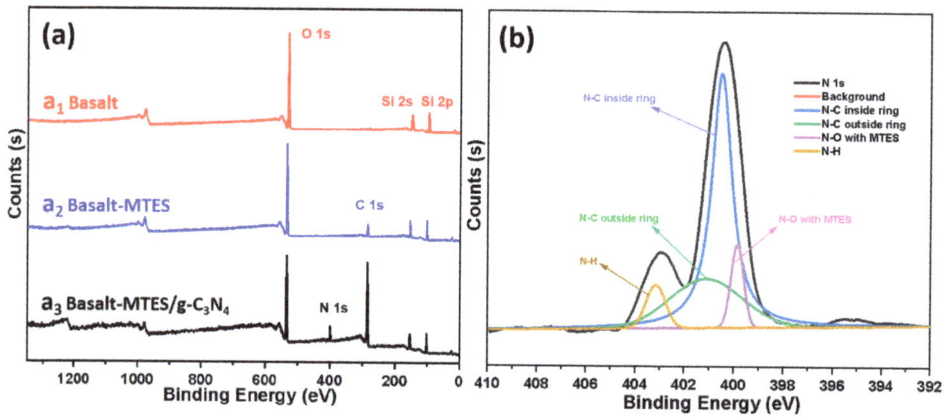

Figure 5. XPS spectra of Basalt, Basalt–MTES, and Basalt–MTES/g-C_3N_4 (**a**); the deconvolution of N 1 s spectra (**b**).

In order to further explore the law of the grafting reaction of melamine on the surface of Basalt, we fitted the N 1s absorption peak of Basalt–MTES/g-C_3N_4. After the grafting reaction, the N in the material may have the following four connection modes, which are the N-C inside the melamine ring, the N-C outside the melamine ring, the N-O connected with the coupling agent, and the N-H not involved in the reaction. Four deconvoluted peaks centered at 400.5 eV, 401.2 eV, 399.9 eV, and 403.2 eV were assigned to the above four connection methods, respectively [33,34]. The N 1s spectra of Basalt–MTES/g-C_3N_4 are illustrated in Figure 5b. According to the fitting results, the peak area ratios of the four groups are 77.12%, 10.51%, 3.85%, and 8.52%, respectively. Among the melamine grafted on the surface of Basalt, the proportion of melamine involved in g-C_3N_4 crystal formation reached 82.63%.

Nitrogen sorption isotherm and pore size distribution of Basalt and Basalt–MTES/g-C_3N_4 are shown in Figure 6. For Basalt (Figure 6(a_1,a_2)), the specific surface area and maximum probability aperture calculated by the BET method and the BJH method are 563.25 $m^2 \cdot g^{-1}$ and 71.27 nm, respectively. After the grafting reaction (Figure 6(b_1,b_2)), the pore size and specific surface area decrease, as shown in Table 1. After the melamine grafting reaction, the material still has a large pore size and specific surface area (406.55 $m^2 \cdot g^{-1}$, 34.29 nm), which can fully ensure that the adsorbed substance can freely enter the pore interior.

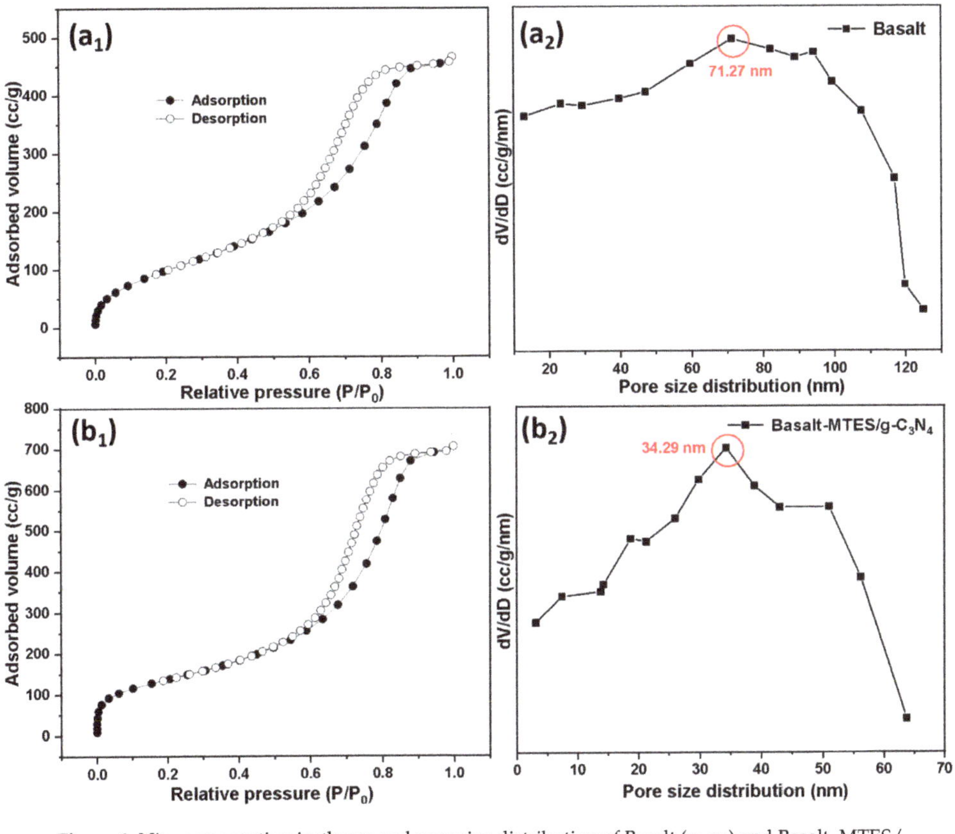

Figure 6. Nitrogen sorption isotherm and pore size distribution of Basalt (**a_1,a_2**) and Basalt–MTES/g-C_3N_4 (**b_1,b_2**).

Table 1. Pore size and specific surface area before and after the grafting reaction.

Species	Pore Size (nm)	Specific Surface Area ($m^2 \cdot g^{-1}$)
Basalt	71.27	563.25
Basalt–MTES/g-C_3N_4	34.29	406.55

2.2. *Adsorption Experiment Results of Cr(VI) by Basalt–MTES/g-C_3N_4*

Figure 7 examines the influence of pH (2–9) on the adsorption of Cr(VI) by three materials: Basalt, Basalt–MTES, and Basalt–MTES/g-C_3N_4. It can be seen from the figure that under any pH conditions, the adsorption effect of Basalt–MTES/g-C_3N_4 on Cr(VI) is much higher than that of Basalt and Basalt–MTES, which indicates that the g-C_3N_4

group has a significant adsorption effect on Cr(VI) [35]. For the adsorption of Cr(VI) by Basalt–MTES/g-C_3N_4 (Figure 7c), with the increase in pH, the adsorption amount first increases and then decreases. The adsorption amount gradually increases within the pH range of 2 to 6, which is because under low pH (2–4) conditions, Cr(VI) mainly exists in the form of $Cr_2O_7{}^{2-}$, and its strong oxidizing property causes a certain degree of damage to the organic functional groups in the material, so the adsorption effect is not ideal. And as the pH increases to 5–6, part of $Cr_2O_7{}^{2-}$ is converted into $CrO_4{}^{2-}$. Meanwhile, the oxidizing power of $CrO_4{}^{2-}$ is limited when the acidity weakens, the damage to the functional groups of the material is reduced, and the adsorption effect is significantly improved. When pH > 6, $Cr_2O_7{}^{2-}$ is converted in large quantities to $CrO_4{}^{2-}$; that is, the concentration of the adsorbate increases, which leads to a certain degree of decline in the adsorption effect within the limited adsorption sites [36]. As the pH continues to rise, the adsorption effect further declines, and the alkalinity of the solution increases. This is because the SiO_2 in the Basalt matrix is corroded under alkaline conditions, resulting in a certain degree of damage to the material framework (pore structure) and affecting the adsorption effect. And as the alkalinity increases, the adsorption capacity decreases, indicating that the material is more damaged. Therefore, pH 5–6 is more conducive to the removal of Cr(VI) by Basalt–MTES/g-C_3N_4.

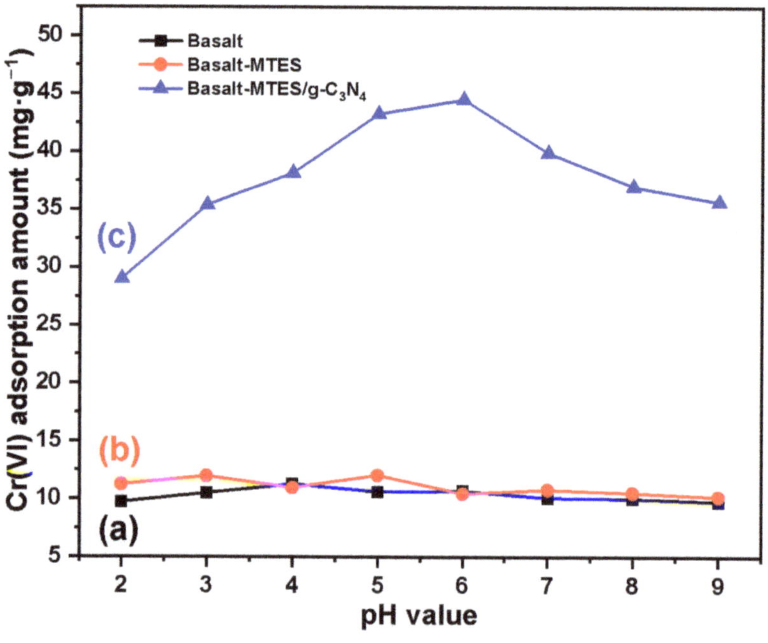

Figure 7. The influence of pH on the adsorption of Cr(VI) by Basalt (**a**), Basalt–MTES (**b**), and Basalt–MTES/g-C_3N_4 (**c**).

The influence of ionic strength on the adsorption of Cr(VI) by Basalt–MTES/g-C_3N_4 is shown in Figure 8. When the Cl^- concentration increased from 0 to 0.1 mol·L^{-1}, the adsorption amount of Cr(VI) decreased significantly; that is, the increase in the Cl^- concentration competed with the Cr(VI) adsorption process, which suggests that the Basalt–MTES/g-C_3N_4 adsorption of Cr(VI) may be an anion exchange process.

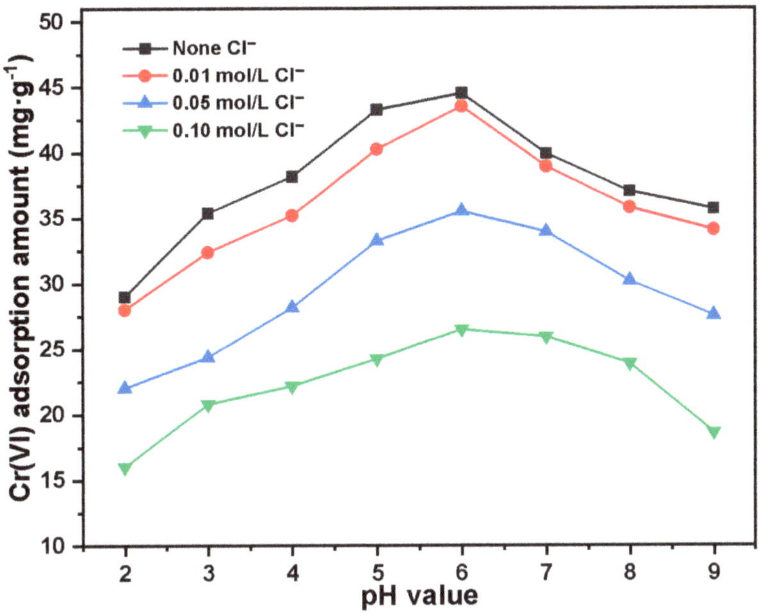

Figure 8. The influence of ionic strength on the adsorption of Cr(VI) by Basalt–MTES/g-C_3N_4.

The adsorption kinetics results of Basalt–MTES/g-C_3N_4 for Cr(VI) are shown in Figure 9. In the variation relationship of 'adsorption capacity–adsorption time', the adsorption of Cr(VI) by the material reached equilibrium within 400 s and remained stable in the remaining contact time. It can be seen from the figure that the adsorption rate was relatively fast, within 0 to 90 s. As time went on, a large number of adsorption sites were occupied, resulting in a slowdown in the adsorption rate. After 400 s, the adsorption tended to be balanced.

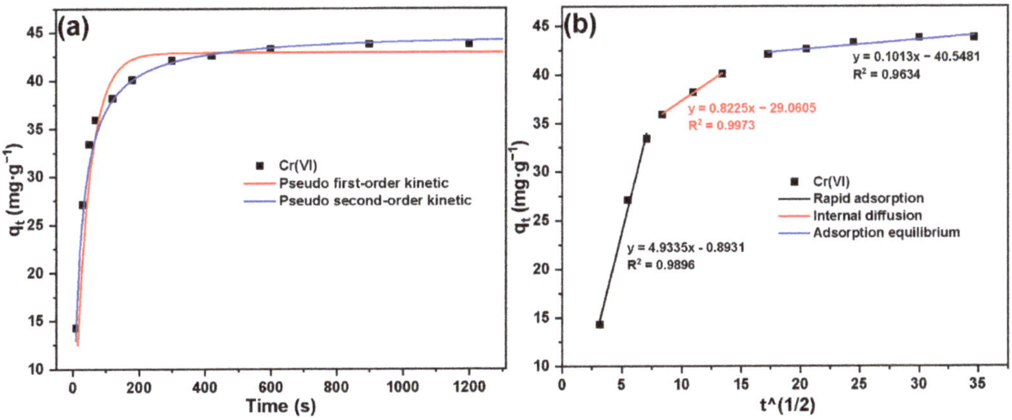

Figure 9. Adsorption kinetic model for Cr(VI) on Basalt–MTES/g-C_3N_4 ((**a**): pseudo-first-order and pseudo-second-order kinetic modes; (**b**): intra-particle diffusion model).

In order to further explore the adsorption mechanism, three dynamic models (pseudo-first-order kinetic model, pseudo-second-order kinetic model, and intra-particle diffusion) were used; the models were fitted to the curve of Basalt–MTES/g-C_3N_4 adsorbing Cr(VI).

The pseudo-first-order kinetic model, pseudo-second-order kinetic model, and intra-particle diffusion model were given as Formulas (1)–(3) [37–39]:

$$q_t = q_e(1 - \exp(-k_1 t)) \tag{1}$$

$$q_t = \frac{q_e^2 k_2 t}{1 + q_e k_2 t} \tag{2}$$

$$q_t = k_p t^{\frac{1}{2}} + C \tag{3}$$

where q_t ($mg \cdot g^{-1}$) is the adsorption capacity at time t (s), and q_e ($mg \cdot g^{-1}$) is the equilibrium adsorption capacity. And k_1 (s^{-1}) is the pseudo-first-order kinetic rate constant, and k_2 ($g \cdot s \cdot mg^{-1}$) is the pseudo-second-order kinetic rate constant. K_p is the intra-particle diffusion rate constant.

The fitting results of the pseudo-first-order kinetic model and pseudo-second-order kinetic model are shown in Figure 9a, and the fitting parameters are shown in Table 2. By comparing the correlation coefficient R^2 and the equilibrium adsorption capacity q_e, it can be known that the correlation coefficient fitted by the pseudo-second-order kinetic model is higher (R^2 = 0.9973), and the calculated theoretical adsorption capacity is more consistent with the experimental data, which indicates that the pseudo-second-order kinetic model can better conform to the process of Basalt–MTES/g-C_3N_4 adsorbing Cr(VI), reflecting that the rate-determining step of this process might be chemical adsorption [40]. Furthermore, the fitting straight line of the intra-particle diffusion model (Figure 9b) does not pass through the origin, indicating that intra-particle diffusion is not the main factor controlling the adsorption rate. The adsorption of Cr(VI) by Basalt–MTES/g-C_3N_4 is mainly based on surface adsorption.

Table 2. Adsorption kinetic parameters (pseudo-first order and pseudo-second order) for Cr(VI) on Basalt–MTES/g-C_3N_4.

Adsorption Kinetic Model	Parameters	Cr(VI)
Pseudo-first order	q_e ($mg \cdot g^{-1}$)	55.28
	K_1 (s^{-1})	0.014
	R^2	0.985
Pseudo-second order	q_e ($mg \cdot g^{-1}$)	42.62
	K_2 ($g \cdot s \cdot mg^{-1}$)	0.086
	R^2	0.997

To determine the interaction between the adsorption material and the adsorbate, we conducted an isotherm study on the adsorption of Cr(VI) by Basalt–MTES/g-C_3N_4, as shown in Figure 10. It can be seen from the figure that the adsorption amount of Cr(VI) increases with the increase in the initial concentration of Cr(VI), which is because the higher concentration greatly increases the contact opportunities between the Cr(VI) ions and the active sites on the surface and inside Basalt–MTES/g-C_3N_4. The adsorption data were fitted using the Langmuir and Freundlich models, respectively, and the formulas are as follows [41,42]:

$$q_e = \frac{q_{max} b C_e}{1 + b C_e} \tag{4}$$

$$q_e = K_f C_e^{n_f} \tag{5}$$

where q_e (mg·g^{-1}) is the equilibrium adsorption capacity, q_{max} (mg·g^{-1}) is the maximum adsorption capacity, C_e (mg·g^{-1}) is the equilibrium concentration, b is the constant, K_f is the Freundlich constant, and n_f is the concentration index.

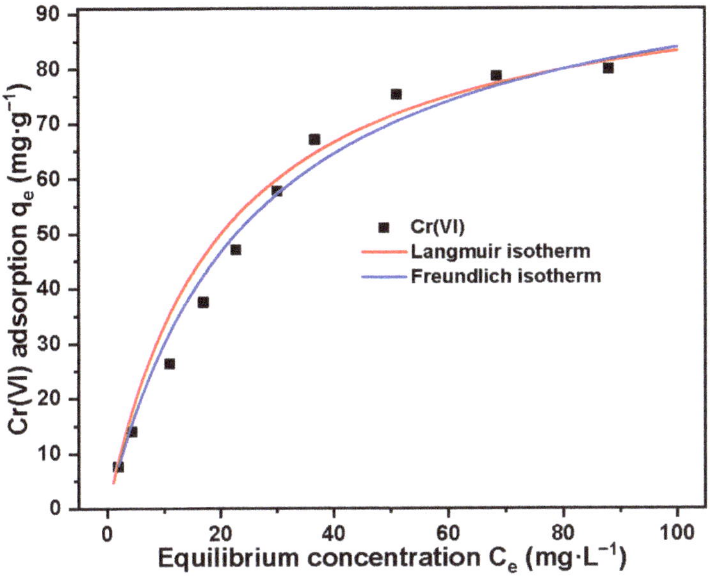

Figure 10. Adsorption isotherm model for Cr(VI) on Basalt–MTES/g-C_3N_4.

The fitting parameters are shown in Table 3 as follows. Through the comparison of the correlation coefficients of the two models, the adsorption process of Cr(VI) by Basalt–MTES/g-C_3N_4 is more in line with the Langmuir model. The maximum adsorption amount calculated by the Langmuir model is close to the experimental data. The adsorption behavior conforms to the monolayer adsorption on heterogeneous surfaces, and the interaction between adjacent $Cr_2O_7^{2-}$ (or CrO_4^{2-}) can be ignored [43]. The maximum adsorption amount calculated by the Langmuir model was 77.26 mg·g^{-1}.

Table 3. Langmuir and Freundlich adsorption isotherm parameters for Cr(VI) on Basalt–MTES/g-C_3N_4.

Adsorption Isotherm Model	Parameters	Cr(VI)
Langmuir	q_{max} (mg·g^{-1})	77.26
	b (L·mg^{-1})	0.093
	R^2	0.992
Freundlich	K_f	8.538
	n_f	0.354
	R^2	0.946

The Van 't Hoff equation was used to fit the adsorption capacity at different temperatures to obtain the thermodynamic parameters ΔH and ΔS. We calculated the ΔG at different temperatures using the Gibbs equation. The Van 't Hoff and Gibbs equations were given as Formulas (6) and (7):

$$\ln K_c = -\left(\frac{\Delta H}{R}\right)\frac{1}{T} + \frac{\Delta S}{R} \quad (6)$$

$$\Delta G = \Delta H - T\Delta S \tag{7}$$

and

$$K_c = \frac{C_s}{C_e} \tag{8}$$

where C_s is the concentration of the solid surface at the adsorption equilibrium, and C_e is the concentration in the solution at the adsorption equilibrium.

Van 't Hoff's fitting results and parameters for temperature changes are shown in Figure 11 and Table 4. According to the calculation results of the thermodynamic parameters, the ΔH of Basalt–MTES/g-C_3N_4 for Cr(VI) adsorption is 31.43 kJ·mol^{-1}, which means that the adsorption processes are endothermic. According to the study of Spek, D et al., the enthalpy change range of physical adsorption is between 2.10 and 20.90 kJ·mol^{-1}, and the enthalpy change range of chemical adsorption is between 20.90 and 418.40 kJ·mol^{-1} [44]. In the temperature range of 293.15–333.15 K, the ΔG range of Basalt–MTES/g-C_3N_4 for Cr(VI) adsorption is between −0.71 and −5.10 kJ·mol^{-1}; namely, with the increase in temperature, the ΔG value decreases, indicating that the adsorption process is spontaneous.

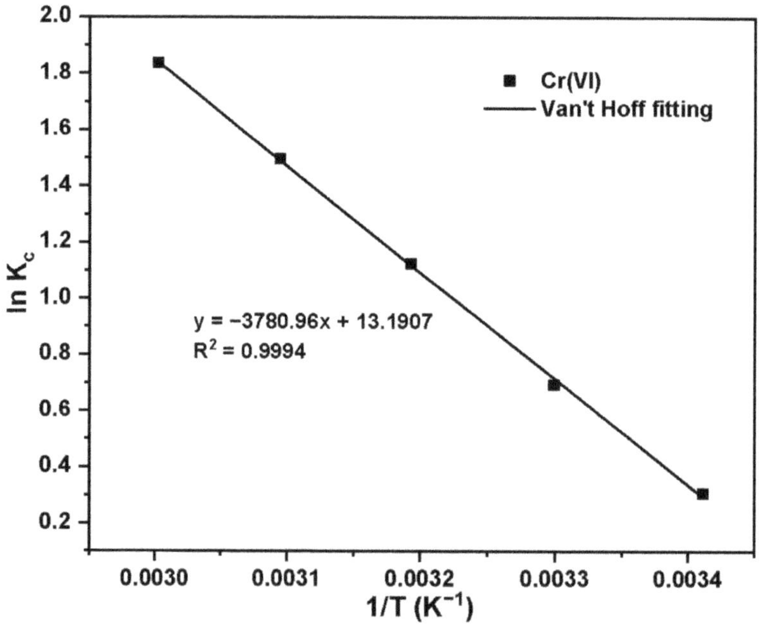

Figure 11. Fitting results of K_c and 1/T of Cr(VI).

Table 4. Adsorption thermodynamic parameters of Cr(VI).

Species	T (K)	ΔG (KJ·mol^{-1})	ΔH (KJ·mol^{-1})	ΔS (J·mol^{-1}·K^{-1})	R^2
Cr(VI)	293.15	−0.71	31.43	109.67	0.9994
	303.15	−1.81			
	313.15	−2.91			
	323.15	−4.02			
	333.15	−5.10			

2.3. *Analytical Pretreatment Experimental Result of Cr(VI) by Basalt–MTES/g-C_3N_4*

The analytical pretreatment results (elution curves) of Cr(VI) in simulated samples and real ethylene wastewater samples by Basalt–MTES/g-C_3N_4 are shown in Figure 12, and the

Cr(VI) recovery rates calculated based on the experiment results are presented in Table 5. According to the calculation results, for the three simulated samples of Cr(VI) with different concentrations (initial concentrations of 0.01 mg·L^{-1}, 0.02 mg·L^{-1}, and 0.03 mg·L^{-1}), the average recovery rates all exceeded 99%. In the tests of the three groups of real wastewater samples, the overall recovery rate was slightly lower than that of the simulated samples, which might be caused by the more complex components in the ethylene wastewater (such as other metal elements and small-molecule organic substances) [45]. However, the recovery rate of Cr in real samples can still reach about 97%. The above experiment results show that for Cr(VI) lower than the detection limit, Basalt–MTES/g-C_3N_4 can effectively enrich it by using the extraction chromatographic layer method, thereby achieving the purpose of an accurate analysis.

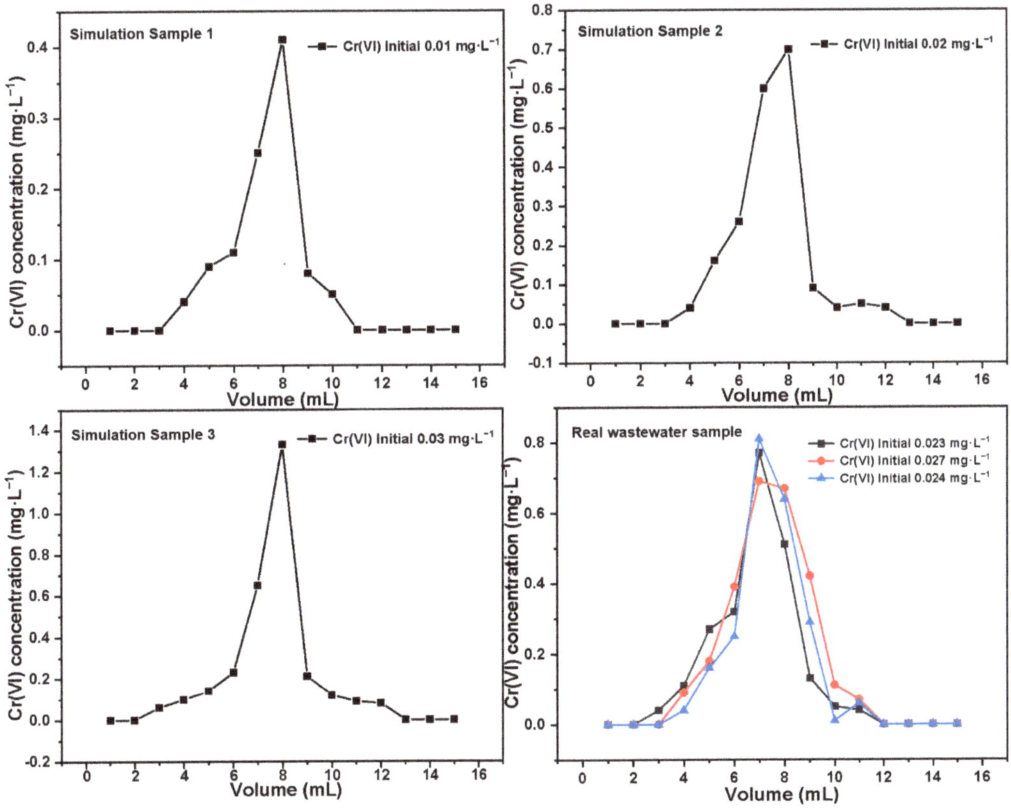

Figure 12. The analytical pretreatment elution curve of Basalt–MTES/g-C_3N_4 for Cr(VI).

Table 5. Statistics on the recovery rate of Cr(VI) in simulated samples and real samples.

Species	Initial Cr(VI) Concentration of the Sample		Enrichment Rate
Cr(VI)	Simulated sample	0.01 mg·L^{-1}	99.21%
		0.02 mg·L^{-1}	99.54%
		0.03 mg·L^{-1}	99.33%
	Real sample	0.023 mg·L^{-1}	97.31%
		0.027 mg·L^{-1}	98.02%
		0.024 mg·L^{-1}	97.24%

2.4. The Result of the Cyclic Regeneration Experiment

The Basalt–MTES/g-C_3N_4 cycling regeneration experiment is shown in Figure 13. After five cycles, the removal rates of Cr(VI) are 96.87%, 96.31%, 95.64%, 95.14%, and 94.27%, respectively. Under the same experimental conditions, the adsorption capacity of the material slightly decreases with the increase in the number of repeated reactions. This might be due to the fact that the potentially adsorbed Cr(VI) cannot be completely desorbed, resulting in a slight reduction in adsorption sites during the repeated adsorption process. The above results indicate that the Basalt–MTES/g-C_3N_4 material has good regeneration ability and reusability and is an effective pretreatment material for Cr(VI) analysis in ethylene wastewater.

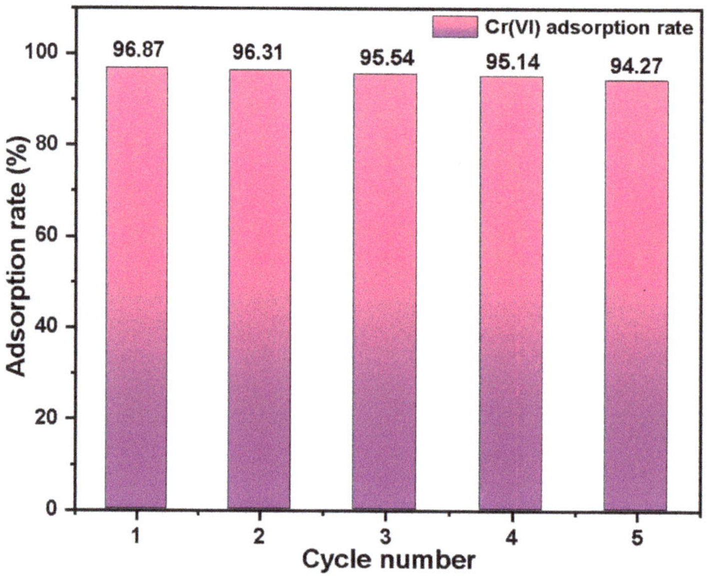

Figure 13. The influence of the number of cycles and regenerations on the removal of Cr(VI) by Basalt–MTES/g-C_3N_4.

3. Materials and Methods

3.1. Materials

Natural Basalt ore (Chengde Zhongxian Technology Co., Ltd., Longgang, China); melamine (AR, Maclin Reagents Shanghai Co., Ltd., Shanghai, China); hydroxymethyl triethoxysilan (MTES) (98%, Maclin Reagents Shanghai Co., Ltd., Shanghai, China); HNO_3 (AR, Maclin Reagents Shanghai Co., Ltd., Shanghai, China); anhydrous ethanol (Maclin Reagents Shanghai Co., Ltd., Shanghai, China); $K_2Cr_2O_7$ solution (1000 mg·L^{-1}, Maclin Reagents Shanghai Co., Ltd., Shanghai, China); ethylene cracking untreated wastewater (Sinopec Group, Yanshan Petrochemical Company, Beijing, China).

3.2. Preparation of Basalt Matrix-Grafted Graphitic Carbon Nitride Material (Basalt–MTES/g-C_3N_4)

3.2.1. Pretreatment of Basalt–MTES

The natural Basalt ore is crushed by a crusher and screened to retain Basalt particles with a size of 80–100 mesh. In the above particles, deionized water is added and then placed in an ultrasonic cleaner for cleaning (20 min) to remove impurities other than the mineral components. After cleaning the Basalt particle material, 2 mol·L^{-1} of HNO_3 is added, and it is stirred with a magnetic stirrer for 60 min. The above particles are washed with deionized water and dried, and the porous Basalt-based material (Basalt) is prepared for use.

The Basalt was modified with the coupling agent hydroxymethyl triethoxysilane (MTES) to obtain alkylated Basalt. Using xylene as the solvent, the Basalt was placed in a reactor and stirred for 30 min, then MTES was added, and the mixture was stirred and refluxed for 24 h at a constant temperature (80 °C). After the grafting reaction, the product was washed three times with ethanol and dried under vacuum at 60 °C for 24 h to obtain modified Basalt (Basalt–MTES).

3.2.2. Preparation of Basalt–MTES/g-C_3N_4

We put 10 g of melamine in an alumina crucible, placed the crucible in a muffle furnace, heated it to 550 °C (2.5 °C/min), and maintained it for 4 h to allow the melamine to react fully. After the reaction was completed, the product was cooled to room temperature in a muffle furnace. We collected the product and grinded it into powder. Then, we rinsed it repeatedly with deionized water and anhydrous ethanol until neutral, and dried it at 70 °C for 24 h. Finally, the dried material was ground and sieved to obtain g-C_3N_4.

Using acetic acid solution with a mass fraction of 2.5% as the solvent, Basalt–MTES was put into the reactor and stirred continuously to keep it in a suspended state. Then, we added the g-C_3N_4 powder to the above-mentioned mixing system and continued to stir at room temperature for 24 h. Then it was left to stand overnight. After filtration, washing, and freeze-drying, the product was obtained, which was Basalt–MTES/g-C_3N_4.

3.3. *Characterization of Basalt–MTES/g-C_3N_4*

The morphologies of Basalt, Basalt–MTES, and Basalt–MTES/g-C_3N_4 were observed and recorded, respectively, by scanning electron microscopy (SEM, JEM-2100F, Tokyo, Japan); the changes in the surface chemical groups of Basalt, Basalt–MTES, and Basalt–MTES/g-C_3N_4 were analyzed by a Fourier Transform Infrared spectrometer (FT-IR, Nicolet-460, Waltham, MA, USA). We mixed Basalt, Basalt–MTES, and Basalt–MTES/g-C_3N_4, respectively, at a ratio of 1:100 evenly, and then placed them on a manual tablet press to make thin slices. The test was conducted under the following conditions: a resolution of 4 cm^{-1}, a scanning range of 4000–650 cm^{-1}, and 16 scans; the XRD spectra of Basalt, Basalt–MTES, and Basalt–MTES/g-C_3N_4 were recorded on a powder diffractometer (D8Quest, Bruker, Zweibrücken, Germany), operating at 40 kV and 40 mA. The elements and chemical states existing in Basalt–MTES/g-C_3N_4 were determined by an X-ray photoelectron spectrometer (XPS, Escalalab 250 XI, Thermo Fisher Scientific, Waltham, MA, USA). The specific surface area and pore structure of Basalt and Basalt–MTES/g-C_3N_4 were measured on a specific surface area and porosity analyzer (JT-2000, Haixinrui, Beijing, China).

3.4. *Calculation Method of Adsorption Experiment*

The total concentrations of Cr(VI) were determined by a flame atomic absorption spectrophotometer (TAS-990, Puxi, Beijing, China). The adsorption amount Q of Cr(VI) in aqueous solution by the Basalt–MTES/g-C_3N_4 was calculated by the following Equation:

$$Q = \frac{(C_0 - C_e) * V}{m} \tag{9}$$

where Q ($mg \cdot g^{-1}$) is the adsorption amount, C_0 ($mg \cdot L^{-1}$) is the initial concentration of Cr(VI) before adsorption, and C_e ($mg \cdot L^{-1}$) is the concentration of Cr(VI) after adsorption. V (L) is the volume of the solution, and m (g) is the mass of the Basalt–MTES/g-C_3N_4.

The removal rate η(%) of Cr(VI) in aqueous solution by the Basalt–MTES/g-C_3N_4 was calculated by the following Equation:

$$\eta = \frac{C_e}{C_0} * 100\% \tag{10}$$

3.5. Adsorption Experiment of Cr(VI) on Basalt–MTES/g-C_3N_4

The influence of Basalt–MTES/g-C_3N_4 on the removal of Cr(VI) under different factors was investigated by a single-factor analysis method to determine the adsorption performance of the material. These experiments mainly include pH and ionic strength, adsorption kinetics research, adsorption isotherm research, and adsorption heat research. After each group of adsorption experiments was completed, the adsorbent was separated by filtration, and the supernatant was taken to determine the concentration of Cr(VI). The adsorption capacity Q_e and the removal rate η of Cr(VI) were calculated using Formulas (9) and (10).

3.6. Analytical Pretreatment Experiment of Cr(VI)

A certain amount of Basalt–MTES/g-C_3N_4 was taken, soaked in deionized water for 24 h, and then filtered by suction. Then, it was washed multiple times with 1 mol·L^{-1} HNO_3 until the washing liquid was colorless and transparent. We washed with deionized water until neutral, and vacuum dried at 80 °C until constant weight was ready for use. The pretreated Basalt–MTES/g-C_3N_4 was loaded into a 5 mL extraction chromatographic column, and the column was passed through with HNO_3 (pH = 5.0) to pre-balance it and set it aside for later use. The natural flow rate of the chromatographic column is approximately 0.5–0.6 mL·min^{-1}. Note: The analytical measurement instrument used in this experiment is a flame atomic absorption spectrophotometer (TAS-990, Puxi, Beijing, China). Experimental verification shows that the detection limit of Cr measurement by this instrument is 0.037 mg·L^{-1}.

Simulation sample experiment: We prepared 100 mL of Cr(VI) solutions with concentrations of 0.01 mg·L^{-1}, 0.02 mg·L^{-1}, and 0.03 mg·L^{-1}, respectively (all concentrations were less than the detection limit of Cr by the instrument) and pH = 5.0. We passed each of the above solutions through the Basalt–MTES/g-C_3N_4 extraction chromatographic column, respectively, then eluted Cr(VI) with a 0.1 mol·L^{-1} $Na_2C_2O_4$ solution, and collected 15 mL of the eluent (collected 15 times, 1 mL each time) [14]. The eluent concentrations of each group were measured, respectively, by a flame atomic absorption spectrophotometer, as mentioned above. Each group of experiments was repeated three times, and the elution curve of Cr(VI) was plotted.

Real wastewater sample experiment: We randomly took three oily water samples from the ground produced by ethylene cracking (with a pH of approximately 5.0). It was known, through concentration analysis, that the initial Cr concentrations were 0.023 mg·L^{-1}, 0.027 mg·L^{-1}, and 0.024 mg·L^{-1}, respectively. In total, 100 mL of a water sample was passed through the Basalt–MTES/g-C_3N_4 extraction chromatographic column, and then Cr(VI) was eluted with a 0.1 mol·L^{-1} $Na_2C_2O_4$ solution [14]. We collected 15 mL of the eluent (collect 15 times, 1 mL each time). We measured the concentration of the eluent and plotted the elution curve of Cr(VI).

3.7. Recycling and Regeneration Experiment

The Basalt–MTES/g-C_3N_4, after adsorbing Cr(VI), was added to a conical flask containing the $Na_2C_2O_4$ solution (0.1 mol·L^{-1}) and desorbed through constant-temperature oscillation for 12 h to dissociate Cr from the surface or interior of the adsorbent. After desorption, we took a certain volume of the supernatant, measured the concentration of Cr(VI), and calculated its removal rate of Cr(VI). The desorbed material was washed with deionized water until neutral, dried, and then repeatedly adsorbed five times under the same conditions.

4. Conclusions

This study successfully prepared a porous Basalt–MTES/g-C_3N_4 composite by grafting g-C_3N_4 onto acid-purified natural Basalt via MTES coupling agents for efficient Cr(VI) enrichment in ethylene wastewater. Characterization revealed uniformly distributed g-C_3N_4 nanoparticles on the Basalt surface, with 82.63% melamine participating in crystallization, while the material exhibited a high specific surface area (403.55 m^2/g) and mesoporous structure (34.29 nm). Adsorption experiments demonstrated that protonated amino groups in g-C_3N_4 under acidic conditions (pH 5–6) dominated Cr(VI) removal through anion exchange. The process followed pseudo-second-order kinetics and Langmuir monolayer adsorption mechanisms. The composite achieved >97% recovery of trace Cr(VI) via chromatographic extraction and maintained 96.87% removal efficiency after five regeneration cycles. This work proposes a cost-effective, recyclable green pretreatment strategy for high-sensitivity Cr(VI) detection in ethylene industrial wastewater, offering dual benefits in environmental remediation and industrial compliance. The design integrates Basalt's structural stability with g-C_3N_4's adsorption functionality, advancing sustainable wastewater treatment technologies.

Author Contributions: Conceptualization, Z.W. and Y.R.; methodology, Z.W. and H.R.; formal analysis, S.L. and J.S.; resources, M.J. and S.H.; data curation, Z.W., Y.R., and J.L.; writing—original draft preparation, Y.R. and H.L.; revising the manuscript, Z.W.; All authors have read and agreed to the published version of the manuscript.

Funding: This work was financially supported by Research on Salt Lake Lithium Extraction Nanofiltration Membrane and Magnesium Lithium Separation Mechanism (ZD2022135).

Institutional Review Board Statement: Not applicable.

Informed Consent Statement: Not applicable.

Data Availability Statement: Please visit the website for relevant data: https://app.globus.org/file-manager?origin_id=82f1b5c6-6e9b-11e5-ba47-22000b92c6ec&origin_path=/tmp/ (accessed on 1 June 2025). After entering, click to select: Data—Grafting Graphite-phase Carbon Nitride Material based on Coupling Agent Hydroxymethyl triethoxysilane Modified Basalt Matrix.

Acknowledgments: Research Project of Hebei Province Higher Education Science.

Conflicts of Interest: The authors declare no conflicts of interest.

References

1. Li, H.; Zhang, S.; Zhou, J.; Wang, Z. Toward carbon-neutral ethylene production: Assessment of the application potential of bio-ethylene production pathways in China. *Biofuels Bioprod. Biorefining* **2022**, *16*, 1568–1582. [CrossRef]
2. Zheng, H.; Chen, J.; Hu, X.; Zhu, F. Biomass production of carbohydrate-rich filamentous microalgae coupled with treatment and nutrients recovery from acrylonitrile butadiene styrene based wastewater: Synergistic enhancement with low carbon dioxide supply strategy. *Bioresour. Technol.* **2022**, *349*, 126829–126842. [CrossRef] [PubMed]
3. Qaretapeh, Z.; Zonoozi, H.; Azarmanesh, R. Biostimulation of styrene-containing wastewater in carrier-supported anaerobic bioreactor towards enhancing biogas production and styrene degradation. *J. Water Process Eng.* **2024**, *64*, 105708–105721. [CrossRef]
4. Zhang, Y.; Wang, X.; Hu, M.; Li, P. Effect of hydraulic retention time (HRT) on the biodegradation of trichloroethylene wastewater and anaerobic bacterial community in the UASB reactor. *Appl. Microbiol. Biotechnol.* **2015**, *99*, 1977–1987. [CrossRef]

5. Fasihi, M.; Rajabi, M.; Barfi, B. Efficacious and environmentally friendly deep eutectic solvent-based liquid-phase microextraction for speciation of Cr(III) and Cr(VI) ions in food and water samples. *Int. J. Environ. Anal. Chem.* **2020**, *102*, 1–13. [CrossRef]
6. Vanda, D.; Atika, S.; Eka, Y. The effect of cetyltrimethylammonium bromide (CTAB) addition on green synthesis of porous N-doped TiO_2 for photoreduction of heavy metal ion Cr(vi). *RSC Adv.* **2023**, *13*, 29645–29656.
7. Stylianos, S.; Konstantinos, S.; Manassis, M. Reductive precipitation and removal of Cr(VI) from groundwaters by pipe flocculation-microfiltration. *Environ. Sci. Pollut. Res. Int.* **2018**, *25*, 12256–12262. [CrossRef]
8. Hu, J.; Li, B.; Su, M. Highly sensitive and selective ratiometric fluorescence and colorimetric detection of Cr(VI) via in situ encapsulation of green/red carbon quantum dots in zeolite. *Colloids Surf. A Physicochem. Eng. Asp.* **2025**, *705*, 135565–135572. [CrossRef]
9. Ali, K.; Setia, W.; Rinda, H. Quantification of the Trace Metal Element Cr in Stainless Steel Using Picosecond Laser-Induced Breakdown Spectroscopy at Atmospheric Pressure. *Arab. J. Sci. Eng.* **2023**, *4*, 8165–8172.
10. Liang, H.; Ding, W.; Zhang, H.; Peng, P. A novel lignin-based hierarchical porous carbon for efficient and selective removal of Cr(VI) from wastewater. *Int. J. Biol. Macromol.* **2022**, *204*, 310–320. [CrossRef]
11. Tao, X.; Hu, X.; Wen, Z. Highly efficient Cr(VI) removal from industrial electroplating wastewater over Bi_2S_3 nanostructures prepared by dual sulfur-precursors: Insights on the promotion effect of sulfate ions. *J. Hazard. Mater.* **2022**, *424*, 127423–127436. [CrossRef] [PubMed]
12. Ferhat, S.; Ounissa, S.; Stéphane, M. Polymer inclusion membranes based on CTA/PBAT blend containing Aliquat 336 as extractant for removal of Cr(VI): Efficiency, stability and selectivity. *React. Funct. Polym.* **2019**, *139*, 120–132.
13. Zhang, Y.; Shi, D.; Xu, G. Efficient reduction of Cr(VI) over nitrogen defected 1D/2D CuO/g-C_3N_4 heterojunction photocatalysts. *J. Porous Mater.* **2024**, *32*, 651–664. [CrossRef]
14. Pattnaik, P.; Mohanty, A.; Parida, K. A timely update on g-C_3N_4-based photocatalysts towards the remediation of Cr(VI) in aqueous streams. *RSC Adv.* **2024**, *14*, 36816–36834. [CrossRef]
15. Özsoy, İ.; Fidan, S.; Bora, Ö.; Ürgün, S. Understanding the Damage Mechanisms of Basalt/Carbon Fiber Hybrid Composites Under Quasi-Static and Dynamic Loadings. *Polymers* **2025**, *17*, 866. [CrossRef]
16. Liu, L.; Xiang, D.; Ma, J. Mechanical, electrical, and damage self-sensing properties of basalt fiber/carbon nanotube/poly (arylene ether nitrile) composites. *J. Mater. Sci. Compos.* **2025**, *6*, 2–9. [CrossRef]
17. Kalbasi, R.; Parishani, P.; Mazaheri, O. Encapsulation of Nickel Nanoparticles and Homopoly(Vinylsulfonic Acid) in Mesoporous Carbon CMK-3 as an Acid–Metal Bifunctional Catalyst for Tandem Reductive Amination. *J. Clust. Sci.* **2018**, *29*, 214–227.
18. Çelikten, S.; Çavuşoğlu, B. Mechanical and microstructural properties of fiber-reinforced basalt rock cutting waste-based geopolymer composites exposed to high temperatures. *Environ. Sci. Pollut. Res. Int.* **2025**, *32*, 3629–3648. [CrossRef] [PubMed]
19. Parvizi, P.; Jalilian, M.; Dearn, D. Comparison of mechanical and physical behaviors of carbon/basalt/epoxy hybrid composite rod with carbon/glass/epoxy hybrid composite rod used in ACCC conductor core. *Polym. Compos.* **2024**, *46*, 5773–5787. [CrossRef]
20. Taha, M.; Alenad, M.; Alhassan, S. Enhanced refractive index and nonlinear optical susceptibility in PVDF-ZrO_2/g-C_3N_4 based nanocomposites for NLO applications. *Opt. Quantum Electron.* **2024**, *56*, 1429–1437. [CrossRef]
21. Cao, Y.; Wang, X.; Ding, L. Thermal and mechanical analysis of thermal break structure in balcony with basalt fiber reinforced polymer. *J. Build. Eng.* **2024**, *98*, 111298–111306. [CrossRef]
22. Nováková, L.; Schnabl, P.; Büchner, J. The characterization of sunburn basalts and their magnetic and petrographic properties. *J. Geosci.* **2018**, *63*, 333–344. [CrossRef]
23. Asfahani, J.; Ghani, A. Automated interpretation of nuclear and electrical well loggings for basalt characterization (case study from southern Syria). *Appl. Radiat. Isot.* **2012**, *70*, 2500–2506. [CrossRef] [PubMed]
24. Purcar, V.; Răditoiu, V.; Dumitru, A. Antireflective coating based on TiO_2 nanoparticles modified with coupling agents via acid-catalyzed sol-gel method. *Appl. Surf. Sci.* **2019**, *4*, 819–824. [CrossRef]
25. Sahebrao, S.; Subash, C. Application of airborne and ground geophysics to unravel the hydrogeological complexity of the Deccan basalts in central India. *Hydrogeol. J.* **2022**, *30*, 2097–2116.
26. Shen, Y.; Liu, Z.; Bi, R.; Zhou, B.; Wang, Y.; Liu, J.; Wang, Z.; Han, B. Low-Crystallized Carbon as an Electron Mediator in g-C_3N_4/C/TiO_2 for Enhancing Photocatalytic Degradation of Antibiotics. *Nanomaterials* **2025**, *15*, 365. [CrossRef] [PubMed]
27. Saeed, S.; Hekmatara, H. MWCNT decorated Magnetic/Ceramic/Polymer (Fe_2O_3/Fe_3O_4/SiO_2/PANI) nanocomposite with enhanced microwave absorption at an entire X and Ku frequency bands. *Ceram. Int.* **2024**, *50*, 11795–11803. [CrossRef]
28. Wu, K.; Song, S.; Wu, H. Carbon quantum dots modified Bi2WO6 nanoflowers for enhancing photocatalytic activity: An experimental and DFT study. *Micro Nano Lett.* **2020**, *15*, 317–322. [CrossRef]
29. Maybodi, S.; Hammood, O. Photodegradation of Polycyclic Aromatic Hydrocarbons Under Visible Light Using Modified g-C_3N_4 as Photocatalyst, Spectroscopic Studies. *Polycycl. Aromat. Compd.* **2025**, *45*, 192–204. [CrossRef]

30. Li, H.; Pang, J.; Jaegers, R. Conversion of ethanol to 1,3–butadiene over Ag–ZrO_2/SiO_2 catalysts: The role of surface interfaces. *J. Energy Chem.* **2021**, *54*, 7–15. [CrossRef]
31. Yuan, Y.; Yin, Y.; Yu, D. Effects of compound coupling agents on the properties of PTFE/SiO_2 microwave composites. *J. Mater. Sci. Mater. Electron.* **2017**, *28*, 3356–3363. [CrossRef]
32. Seddigi, S.; Baig, U.; Khan, A. Visible Light Assisted CuO/g-C_3N_4 Nanocomposite Photocatalyst for Dye Degradation and In-Vitro Deactivation of Pseudomonas aeruginosa Bacteria. *Arab. J. Sci. Eng.* **2024**, *50*, 3925–3936. [CrossRef]
33. Zhao, M.; Zhao, Y.; Wang, C. Studies on ZnO/g-C_3N_4 Heterojunction Photocatalyst for RhB Degradation Under Vis-Light in a MNB-Agitated Pilot Reactor. *Arab. J. Sci. Eng.* **2025**, *50*, 3751–3762. [CrossRef]
34. Murugavel, K.; Vasanthakumar, V.; Pazhanivel, T. Exploring prompt photocatalytic degradation of MB dye using Cu 0.5 Co 0.5 WO_4/g-C_3N_4 nanocomposite under visible light irradiation. *Emergent Mater.* **2024**, *7*, 987–998.
35. Zhong, Q.; Lan, H.; Zhang, M. Preparation of heterostructure g-C_3N_4/ZnO nanorods for high photocatalytic activity on different pollutants (MB, RhB, Cr(VI) and eosin). *Ceram. Int.* **2020**, *46*, 12192–12199. [CrossRef]
36. Yu, C. Removal of Cr(VI) from aqueous solutions by nZVI-loaded sludge-derived biochar: Performance and mechanism. *Water Sci. Technol.* **2022**, *86*, 2089–2105. [CrossRef] [PubMed]
37. Al-Shannag, M.; Al-Qodah, Z.; Bani-Melhem, K. Heavy metal ions removal from metal plating wastewater using electrocoagulation: Kinetic study and process performance. *Chem. Eng. J.* **2015**, *260*, 749–756. [CrossRef]
38. Örnek, A.; Özacar, M.; Şengil, A. Adsorption of lead onto formaldehyde or sulphuric acid treated acorn waste: Equilibrium and kinetic studies. *Biochem. Eng. J.* **2007**, *37*, 192–200. [CrossRef]
39. Yang, W.; Cloete, S.; Morud, J. An Effective Reaction Rate Model for Gas-Solid Reactions with High Intra-Particle Diffusion Resistance. *Int. J. Chem. React. Eng.* **2016**, *14*, 331–342. [CrossRef]
40. Lobo, C.; Castellari, J.; Lerner, C. Functional iron chitosan microspheres synthesized by ionotropic gelation for the removal of arsenic (V) from water. *Int. J. Biol. Macromol.* **2020**, *164*, 1575–1583. [CrossRef]
41. Munawar, I.; Muhammad, S.; Zahid, A. Paracetamol and amoxicillin adsorptive removal from aqueous solution using phosphoric acid activated-carbon. *Z. Phys. Chem.* **2023**, *237*, 257–271.
42. Mitra, A.; Mehran, H.; Arezoo, T. Chromium (VI) bioremoval from contaminated wastewater using Pseudomonas aeruginosa ATHA23 producing biofilm supported on clinoptilolite. *Environ. Geochem. Health* **2022**, *45*, 427–442.
43. Khayyun, S.; Mseer, H. Comparison of the experimental results with the Langmuir and Freundlich models for copper removal on limestone adsorbent. *Appl. Water Sci.* **2019**, *9*, 170. [CrossRef]
44. Spek, D. Selling a Theory: The Role of Molecular Models in J. H. van 't Hoff's Stereochemistry Theory. *Ann. Sci.* **2006**, *63*, 157–177. [CrossRef]
45. Zhong, L.; Ren, W.; Guo, W. A pilot scale test of ozonization treatment of ethene wastewater for reuse. *Front. Chem. Eng. China* **2008**, *2*, 191–195. [CrossRef]

minerals

MDPI

Article

Assessment of Heavy Metal Content and Identification of Their Sources in Bottom Sediments and Various Macrophyte Species of the Narew River (Poland)

Mirosław Skorbiłowicz [1] and Marcin Sidoruk [2,*]

[1] Department of Technology in Environmental Engineering, Faculty of Civil Engineering and Environmental Sciences, Bialystok University of Technology, Wiejska 45E, 15-351 Białystok, Poland; m.skorbilowicz@pb.edu.pl

[2] Department of Water Resources and Climatology, Faculty of Agriculture and Forestry, University of Warmia and Mazury in Olsztyn, Plac Łódzki 2, 10-719 Olsztyn, Poland

* Correspondence: marcin.sidoruk@uwm.edu.pl; Tel.: +48-895234351

Abstract: The condition of the aquatic environment, particularly in protected areas of high ecological value such as the Narew River, requires detailed monitoring to identify and minimise the impact of anthropogenic factors on the ecosystem. This study focused on the content of heavy metals in bottom sediments and macrophytes of the Narew River, emphasising the influence of human activities and natural factors on this ecologically valuable ecosystem. Pb, Cr, Zn, Cd, Fe, and Mn concentrations were analysed in sediment samples, and ten macrophyte species were collected at 11 sampling points along the river. A geochemical index (Igeo) and multivariate statistical analyses were employed to identify sources of contamination. The digested samples (sediments and plants) were analysed for Pb, Cr, Cu, Zn, Ni, Cd, Fe, and Mn using flame atomic absorption spectrometry (AAS) on an ICE 3500 Thermo Scientific spectrometer, with a measurement error below 5%, validated against certified reference materials. The study results indicated that most metals, including Ni, Cr, Co, Fe, and Mn, predominantly originate from natural geological processes. In contrast, Zn, Cd, Cu, and Pb were identified as being enriched due to anthropogenic activities. An analysis of macrophytes revealed varied patterns of metal accumulation, which correspond to the bioavailability of metals and their environmental concentrations. Comprehensive statistical analyses provided insights into the predominant sources of metal contamination, closely associated with industrial emissions, agricultural runoff, and transportation activities. The integration of sediment and macrophyte monitoring allowed for a thorough evaluation of the Narew River ecosystem, facilitating the identification of key pollution sources. These findings highlight the critical need for measures to mitigate anthropogenic contributions of heavy metals—particularly from industrial, agricultural, and transportation sectors—to safeguard the Narew River's unique ecological and natural heritage.

Keywords: heavy metals; river; bottom sediments; macrophytes

1. Introduction

Water quality, particularly in ecosystems of significant rivers and their tributaries, is significantly influenced by human activities within the river and across its catchment area [1]. Rivers transport characteristic pollutants that are accumulated within their basins, serving as critical pathways for the migration and transformation of trace elements and as primary sites for their accumulation [2,3]. Metal pollution in aquatic ecosystems has become a global concern [4,5] due to these elements' high toxicity, persistence, and bioaccumulation potential, which have drawn worldwide attention [6].

The concentrations and bioavailability of trace metals in rivers are affected by numerous factors, including chemical speciation; mobility within aquatic systems; climatic, morphological, and hydrological conditions; and the geological characteristics of the catchment area [7,8]. Organic matter in the environment may also be important in controlling bioavailability through chelation reactions that enhance metal solubility [9]. Additionally, water and sediments' pH and redox potential (Eh) are critical in determining metal bioavailability [10]. Compared to lithogenic metals, anthropogenic metals are more bioavailable and transportable within aquatic environments [11]. Anthropogenic activities such as fertilisation, industrial and municipal wastewater discharge, mining, transportation, stormwater, and atmospheric deposition are mainly responsible for the increasing accumulation of heavy metals in water, sediments, and macrophytes, significantly contributing to aquatic ecosystem pollution [12,13].

Macrophytes can absorb large quantities of metals from their surrounding water, accumulating pollutants throughout their growth period that are unaffected by temporal fluctuations [14]. This makes them excellent indicators of heavy metal contamination in aquatic ecosystems [15,16]. Plant species growing along stream banks and beds absorb and accumulate essential elements, primarily from the water, thereby contributing to protection against eutrophication [17,18]. The choice of appropriate macrophytes depends on species, availability, and climatic conditions.

Heavy metal accumulation in sediments threatens aquatic environments for a long time [19]. Milenkovic et al. [20] argue that sediments can record temporal changes in trace elements caused by environmental changes and human activities. According to Kuang et al. [21], Calmuc et al. [22], and Zonta et al. [23], sediments are effective indicators for monitoring pollution, serving not only as carriers of trace elements but also as potential secondary sources in aquatic environments. In recent decades, various sediment quality indices, including the geoaccumulation index (Igeo), have been developed to provide a straightforward comparative assessment of heavy metal contamination in sediments [24].

The objective of this study was to analyse the variability of heavy metal concentrations in bottom sediments and ten macrophyte species collected from the Narew River, the fifth-longest river in Poland. Large sections of the Narew River are protected within national parks, landscape parks, and Natura 2000 areas. The river retains a natural character throughout its length, featuring numerous floodplains and meanders. It holds substantial ecological value, with some areas designated as wetlands of international importance. Consequently, monitoring its environmental condition and identifying its current and potential pollution sources are a top priority for addressing harmful microelements such as heavy metals.

Environmental samples were collected at 11 sites along the river for analysis. Concentrations of Pb, Cr, Cu, Zn, Ni, Cd, Fe, and Mn were measured in the samples. The primary objectives of this research included (1) analysing heavy metal concentrations in the river's bottom sediments and assessing their pollution levels using the geoaccumulation index (Igeo), (2) examining and comparing metal accumulation in ten macrophyte species growing along the riverbanks, and (3) identifying potential sources and factors influencing heavy metal content in the Narew River through multivariate statistical analysis, cluster analysis (CA), and correlation analysis.

This study contributes to understanding aquatic plants in lowland rivers with natural characteristics, including extensive floodplains and meandering of high ecological value. Furthermore, it sheds light on these plants' ability to accumulate trace elements like sediments, establishing a link between trace element concentrations in plant species and their surrounding environment. The findings provide valuable insights for assessing the environmental risks of heavy metals, identifying anthropogenic impacts, and implementing stricter monitoring of the studied aquatic ecosystem.

2. Materials and Methods

2.1. Study Area

The total length of the Narew River (Figure 1) is 484 km, of which the majority, 448 km, lies within Poland, with a smaller segment of 36 km located in Belarus, where the river originates. The Narew River begins at an elevation of 159 m above sea level in the Orłowo wetland complex, on the edge of the Białowieża Forest, initially known as the "Skaronowski Rów". For only a 1 km stretch, the Narew is the border between Poland and Belarus.

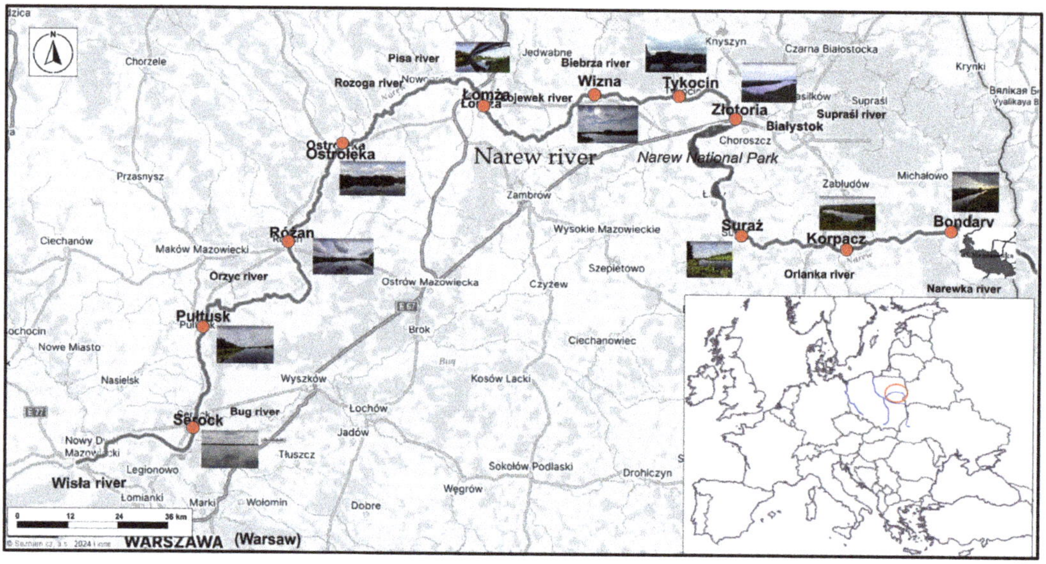

Figure 1. Locations of measurement points along the Narew River in Poland.

This study covered a 431 km section of the river (excluding the Zegrzyński Reservoir). The river basin, characterised by lowland marshy terrain, encompasses an area of 355 km^2 and exhibits a slow flow rate of 0.2–0.3 m/s. The catchment area is 35% forested, while 20% is designated for agricultural use.

The Narew River is connected to a network of tributaries, the most significant of which are the Narewka, Orlanka, Jaskranka, Nereśl, and, particularly, the Supraśl with the Sokołda. The peripheral border position of the Biebrza River to the north of the region and the Bug River to the south limits their contribution to the region's water management. The Narew stretches approximately 210 km from the mouth of the Biebrza River to its confluence with the Bug River within the Zegrzyński Reservoir. Along this stretch, the valley's width ranges from 1.5 km to about 7 km, covering an area of approximately 900 km^2. The catchment area predominantly comprises luvisols and leached brown soils, with podzolic soils, including rusty and podzolic types derived from sands of various origins, ranking third in prevalence.

The climate in the upper part of the Narew River is temperate, influenced by its geographical location, but it also exhibits some effects of a continental climate. Multi-year studies have determined that the average annual temperature is approximately 6.7 °C. The winter season typically begins in late November and ends in early April, with temperatures ranging from −6.7 to −2.7 °C. Subzero temperatures persist for around 70 days annually in this area, compared to approximately 40 days in central Poland.

The Narew River valley includes the city of Ostrołęka, with a population of about 52,000. The city is home to a pulp mill, a 720 MW combined heat and power plant, and other industrial facilities, including those in the metal, construction, and food industries. Below Serock, the lower Narew is crossed by road bridges at Wierzbica, Pułtusk, Różan, Ostrołęka, Nowogród, and Łomża. Among the cities in the region, Łomża is the largest, with a population of around 61,000. Łomża hosts significant industrial facilities, such as a furniture factory, brewery, starch production plant, and cotton industry operations. Further south in the region, the town of Wyszków (approximately 25,000 residents) lies on a railway line from Tłuszcz to Ostrołęka. Wyszków also houses a furniture factory, glassworks, brewery, and other industrial establishments.

The region's industrial production prominently features agri–food processing, with companies such as P.P.S. Łomża and dairy plants in Grajewo, Wysokie Mazowieckie, Kolno, Zambrów, and Piątnica. The textile industry, concentrated in Łomża and Zambrów, and the wood industry in Grajewo also play significant roles.

The water quality of the Narew River is influenced by wastewater discharges from cities such as Łapy, Łomża, Ostrołęka, Różan, and Pułtusk, as well as inflows from tributaries, including the Narewka, Orlanka, Supraśl (carrying wastewater from Białystok), Biebrza, Łojewek, Gać, Pisa, Szkwa, Rozoga, and Orzyc. Transportation routes crossing the river also impact water quality. These include the Hajnówka and Siemianówka railway lines, two road connections between Białystok and Hajnówka, a heavily trafficked road and railway route between Białystok and Bielsk Podlaski, the Łapy–Białystok road, and the Warsaw–Białystok railway line, which crosses the Narew River at two points.

2.2. Sample Collection and Preparation

Macrophyte and bottom sediment samples were collected in August 2020 from 11 measurement points along the Narew River. These measurement points were located near the following towns: Bondary (1), Korpacz (2—downstream of the Orlanka tributary), Suraż (3), Złotoria (4—downstream of the Supraśl tributary), Tykocin (5—downstream of the Nereśl tributary), Wizna (6—downstream of the Biebrza tributary), Łomża (7), Ostrołęka (8), Różan (9), Pułtusk (10), and Serock (11—downstream of the Bug tributary).

Bottom sediments were collected in the coastal zone in the top layer to a depth of 5.0 cm because this layer is where the most intensive deposition of suspended material occurs. A core collector was used to collect samples, which allows for material to be obtained in a way that minimises contamination and sample loss. Ten samples were collected and placed in plastic containers at each selected point. After mixing, a representative sample weighing approximately 1000 g was obtained. These samples were dried and then sieved through a nylon sieve with a mesh diameter of 0.2 mm. The selected fraction was analysed in the laboratory because it constitutes the majority of the bottom sediment composition and plays a key role in transporting substances in the aquatic environment. Sediment samples were collected at a distance of 2–3 m from the macrophyte sampling sites, which ensured data consistency between the sediment and plant analyses.

The following macrophytes, available in the field, were collected for the study: Common reed (Phragmites australis), Reed canary grass (*Phalaris arundinacea* L.), Long-leaf pondweed (*Potamogeton nodosus* Poin.), Clasping-leaf pondweed (*Potamogeton perfoliatus* L.), Sharp-leaf pondweed (*Potamogeton acruifolius*), Floating pondweed (*Potamogeton natans* L.), Slender sedge (*Carex acuta* L.), Yellow water-lily (*Nuphar lutea* L.), Branched bur-reed (*Sparganium erectum* L.), and Sweet manna grass (*Glyceria maxima* Hartman).

Plant material was harvested in August, during the peak of the vegetation season when bioaccumulation is most intense, as noted by Ruiz et al. [25] and Eid et al. [26]. Samples were collected on climatically similar days to minimise the influence of weather conditions. Randomly, four to six specimens of each available species were taken from each sampling point and combined into composite samples. The collected macrophytes were thoroughly washed with river water to remove attached sediments and epiphytes. For bioaccumulation assessment, the macrophytes were placed in labelled polyethene zip-lock bags. Sediment samples (<0.2 mm fraction) and macrophyte samples were dried in the laboratory to constant weight and then ground into a uniform powder using an agate mortar.

The prepared material was transferred to Teflon vessels, and a mixture of 8 mL of HNO_3 and 2 mL of 30% H_2O_2 was added. According to the manufacturer's guidelines, the samples were digested using a microwave digestion system (Ethos Easy, Milestone, Italy). Du Laing et al. [27] reported that microwave-assisted digestion with a mixture of HNO_3 and H_2O_2 is the most appropriate method for determining heavy metals in macrophytes and sediments.

The digested samples (sediments and plants) were analysed for the concentrations of Pb, Cr, Cu, Zn, Ni, Cd, Fe, and Mn using flame atomic absorption spectrometry (AAS) on an ICE 3500 Thermo Scientific spectrometer.

Measurement error was estimated by comparing the results with certified reference materials, including ERM-CD281 grass mixture and LGC7162 strawberry leaves. The calculated measurement error was at most 5% of the certified values. Sediment results were validated using the certified reference material for sediments, NCS DC 73317a. For multivariate statistical analyses, arithmetic means of metal concentrations in different macrophyte organs were calculated and treated as representative of the overall metal content in macrophytes. This approach ensured reliable data for further evaluation of heavy metal bioaccumulation and sediment contamination.

2.3. *Assessment of Sediment Contamination, Geoaccumulation Index (Igeo)*

Müller [28] introduced the geoaccumulation index (Igeo) in 1969 to assess soil contamination, and it has since become a widely used tool for evaluating soil pollution levels [29–32]. The Igeo determines metal enrichment by comparing the concentration of each studied element with its baseline value in the Earth's crust [33].

The local geochemical background (LGB) represents the natural concentrations of trace elements in the environment, shaped by the region's specific geological features. LGB values help assess local pollution levels by reflecting the area's unique geochemical characteristics. This study obtained LGB values from the scientific literature and regional geological data, including information on source rocks and sediments in the Narew region.

However, due to incomplete data on some elements and to ensure comparability with other studies, the global geochemical background (GCB) values proposed by Turekian and Wedepohl [34] were used to calculate the Igeo. These values are widely accepted in environmental studies, facilitating comparisons across geographical areas. The Igeo is defined using the following formula [28]:

$$I_{geo} = \log_2\left(\frac{C_n}{1.5 \cdot GCB}\right)$$

where Cn represents the measured concentration of the studied metal in bottom sediments, GCB denotes the geochemical background content based on the classification proposed by Turekian and Wedepohl [34], and 1.5 is a constant correction factor accounting for the variability of the lithogenic background matrix [28,29,35].

The geoaccumulation index (Igeo) is divided into seven classes: unpolluted, class 0 (Igeo $\leq$ 0); unpolluted or moderately polluted, class 1 (0 < Igeo < 1); moderately polluted, class 2 (1 < Igeo < 2); moderately or heavily polluted, class 3 (2 < Igeo < 3); heavily polluted, class 4 (3 < Igeo < 4); heavily to very heavily polluted, class 5 (4 < Igeo < 5); very heavily polluted, class 6 (Igeo $\geq$ 5) [28]. Turekian and Wedepohl [34] proposed the following geochemical background values: Cr = 90 mg/kg, Zn = 95 mg/kg, Cd = 0.3 mg/kg, Co = 19 mg/kg, Mn = 850 mg/kg, Cu = 45 mg/kg, Ni = 68 mg/kg, Pb = 20 mg/kg and Fe = 47,700 mg/kg.

2.4. Statistical Analysis

The obtained data were analysed using STATISTICA software (version 13.3) before performing analyses, and variables and cases were prepared and organised in a necessary database of research results. The analyses began with the calculation of descriptive statistics, including arithmetic mean, median, maximum, minimum, standard deviation (SD), and coefficient of variation (CV [%]). The normality of the data distribution was verified using the Shapiro–Wilk test, with statistical significance set at $p < 0.05$. Pearson's correlation analysis was employed to investigate relationships between pairs of metals. Pearson's correlation coefficients between various elements provide valuable insights into their sources in the geochemical environment [36].

Hierarchical cluster analysis (HCA) using Ward's method and Factor Analysis (FA) was applied to identify the sources of the studied metals. Before conducting FA, the Kaiser–Meyer–Olkin (KMO) test and Bartlett's test of sphericity were performed. The KMO values ranged from 0.58 to 0.63, and Bartlett's test showed statistical significance. The Varimax rotation method was employed. For interpreting the Factor Analysis results, associations between the original variables and factors were considered significant when the absolute values of their loadings exceeded 0.70. Multivariate statistical methods are widely used for identifying numerous substances and pollutants in the environment and exploring their migration pathways, as demonstrated by Lu et al. [37], Chen et al. [38], Caeiro et al. [39], and Lee et al. [40].

3. Results and Discussion

3.1. Metal Content of Sediment

The results of basic statistical calculations for sediment studies (Table 1) and macrophytes (Table 2) are presented below. The variability of metals at the studied locations was assessed using standard deviation (SD) and the coefficient of variation (CV). The variability of metal concentrations in sediments ranged from a CV of 25.96% (Mn) to 78.42% (Cd). In contrast, the variability of macrophyte metals ranged from a CV of 15.52% (Mn) to 80.40% (Cd). Interestingly, these extreme ranges pertain to the same metals (Mn and Cd) in sediments and macrophytes.

Table 1. Content of analysed metals in the bottom sediments of the Narew River [mg/kg].

Intake Point	Metal								
	Ni	Cr	Zn	Cd	Co	Cu	Pb	Fe	Mn
Bondary	13.98	14.17	151	0.93	7.94	48.01	21.06	12,505	275
Korpacz	2.04	2.80	51	0.02	2.25	15.31	12.18	7529	276
Suraż	5.72	4.46	53	0.23	4.12	18.97	12.94	5980	235
Złotoria	3.31	15.75	147	0.31	3.39	19.36	24.90	11,467	332
Tykocin	3.86	8.74	182	0.35	4.42	20.28	42.90	4737	232
Wizna	6.47	10.46	85	0.50	5.57	16.56	18.98	7110	315
Łomża	6.04	12.18	180	0.65	5.57	19.82	15.73	10,032	425

Table 1. *Cont.*

Intake Point	Ni	Cr	Zn	Cd	Metal Co	Cu	Pb	Fe	Mn
Ostrołęka	5.09	4.70	72	0.23	3.05	15.86	12.85	5039	217
Różan	13.01	11.08	129	0.45	4.94	29.50	17.46	12,704	372
Pułtusk	5.57	3.91	84	0.17	3.60	18.29	11.60	4735	277
Serock	14.87	0.91	33	0.01	1.89	16.90	7.63	2222	167
GCB	68	90	95	0.3	19	45	20	47,200	850
LGB	5	5	48	0.5	2	6	10	n.a.	n.a.
average	7.27	8.10	106	0.35	4.25	21.71	18.02	7642	284
min	2.04	0.91	33	0.01	1.89	15.31	7.63	2222	167
max	14.87	15.75	182	0.93	7.94	48.01	42.90	12,704	425
SD	4.50	4.99	54	0.27	1.73	9.53	9.58	3544	74
CV [%]	61.93	61.58	51	78.42	40.60	43.88	53.16	46	26

Note: SD—standard deviation, LGB—local geochemical background, GCB—global geochemical background, n.a.—not available.

Table 2. Content of analysed metals in submerged parts of macrophytes [mg/kg].

Intake Point	Macrophyte Species	Ni	Cr	Zn	Cd	Co	Cu	Pb	Fe	Mn
Bondary	*Phragmites australis*	6.46	3.36	59	0.12	3.27	24.20	12.28	6211	259
	Potamogeton nodosus Poin.	10.42	0.30	50	0.19	2.61	13.01	9.66	1787	320
Korpacz	*Phalaris arundinacea* L.	1.83	1.37	86	0.10	2.55	17.29	10.33	7797	373
	Potamogeton perfoliatus L.	7.65	3.72	68	0.86	6.51	18.87	11.99	7452	490
	Potamogeton acruifoliud	6.56	3.67	83	0.32	5.12	16.44	9.88	13,127	493
Suraż	*Phragmites australis*	7.36	0.99	67	0.07	2.17	14.94	9.09	6236	428
Złotoria	*Phragmites australis*	5.51	1.61	93	0.07	2.29	13.25	6.67	3980	421
Tykocin	*Phragmites australis*	6.52	1.91	55	0.26	3.27	17.68	8.43	6813	418
	Potamogeton natans L.	11.35	4.18	104	0.44	4.52	17.92	9.02	6738	491
Wizna	*Carex acuta* L.	3.91	2.81	78	0.30	2.63	19.36	7.40	3605	486
	Potamogeton nodosus Poin.	7.02	5.83	108	0.36	4.00	20.90	8.45	6040	488
Łomża	*Phalaris arundinacea* L.	7.91	2.34	61	0.22	3.08	13.45	6.10	6455	459
Ostrołęka	*Potamogeton nodosus* Poin.	9.54	3.15	83	0.28	4.27	27.67	8.15	7888	485
	Carex acuta L.	6.26	3.41	91	0.14	3.00	17.92	7.57	7637	402
Różan	*Phalaris arundinacea* L.	4.23	2.68	72	0.22	2.82	23.02	7.57	7929	338
	Nuphary lutea L.	4.97	3.87	103	0.25	3.21	22.34	8.87	5495	466
Pułtusk	*Nuphary lutea* L.	5.84	2.02	67	0.03	2.73	11.46	6.53	7185	414
	Sparganium erectum L.	1.93	0.90	51	0.09	1.86	9.45	6.01	3632	435
Serock	*Phalaris arundinacea* L.	1.30	1.18	54	0.15	2.21	11.91	6.20	5126	341
	Glyceria maxima Hartman	7.17	2.49	58	0.13	2.99	21.19	8.05	6300	470
	Nuphary lutea L.	5.44	1.54	54	0.12	2.54	14.60	7.40	2293	467
average		6.15	2.54	74	0.22	3.22	17.47	8.37	6178	426
min		1.30	0.30	50	0.03	1.86	9.45	6.01	1787	259
max		11.35	5.83	108	0.86	6.51	27.67	12.28	13,127	493
SD		2.62	1.34	19	0.18	1.11	4.67	1.77	2412	66
CV [%]		42.54	52.65	25	80.40	34.58	26.73	21.11	39	16
Natural content		0.1–5 a	0.02–0.5 a	10–70 a	0.05–0.2 a	0.01–0.8	5–30 a	0.1–5 b	50–200 b	20–500 a

a—Kabata-Pendias and Pendias [41], b—Markert [42].

At the primary statistical analysis stage, it is possible to identify, to a limited extent, the sources of metals in the aquatic environment. Additional statistical tools were applied to expand the scope of the analysis. The mean metal concentrations in the sediments of the Narew River decreased in the following order: Fe > Mn > Zn > Cu > Pb > Cr > Ni > Co > Cd.

To assess the contamination of bottom sediments, the geochemical background established by Turekian and Wedepohl [34] was utilised. The analyses revealed that Ni, Cr, Co, Fe, and Mn concentrations were within natural levels, while Zn, Cd, and, partially, Cu and Pb exceed these levels. The highest concentrations in the sediments were observed for Fe, Mn, and Zn, and the lowest for Cd. This distribution of the studied components is related to their geochemical properties and the degree of their prevalence in the environment. The concentrations of Fe and Mn ranged from 2222 to 12,704 mg/kg and 167 to 425 mg/kg, respectively.

For Zn and Cu, the concentrations ranged from 33 to 182 mgZn/kg and 15.31 to 48.01 mgCu/kg with mean values of 106 ± 54 mgZn/kg and 21.71 ± 9.53 mgCu/kg. This study demonstrated relatively low Pb, Cr, Ni, and Co concentrations, which fell within the range of 7.63–42.90, 0.91–15.75, 2.04–14.87, and 1.89–7.94 mg/kg, respectively. These ranges mostly align with natural levels, except Pb, whose upper range exceeds natural levels by more than two times. The maximum concentration of Cd was approximately three times higher than the geochemical background, suggesting the anthropogenic enrichment of sediments with this element [9,33]. The average Cd concentration in the Narew sediments was 0.35 ± 0.27 mg/kg, ranging from 0.01 to 0.93 mg/kg.

The sediments' Cr, Cd, Cu, Zn, and Co concentrations are strongly intercorrelated, as indicated by correlation coefficients ranging from r = 0.61 to r = 97. According to Ke et al. [43], Bhuyan et al. [44], and Ciazela et al. [45], consistent and robust correlations between pairs of metals, with values between 0.7 and 0.9, suggest a common origin.

The geochemical index values (Igeo) for Ni, Co, Cu, Fe, Mn, and Cr (Table 3) were negative. Consequently, most of the studied sediments were classified as class 0, indicating unpolluted. However, the Igeo values for Cd, Pb, and Zn were positive in specific locations.

Table 3. Geochemical index of bottom sediments River Narew.

Measurement	Point	Ni	Co	Pb	Cu	Zn	Fe	Mn	Cr	Cd
Bondary	1	−2.87	−1.84	−0.51	−0.49	0.08	−2.52	−2.21	−3.25	1.05
Korpacz	2	−5.65	−3.66	−1.30	−2.14	−1.48	−3.25	−2.21	−5.59	−4.72
Suraz	3	−4.16	−2.79	−1.21	−1.83	−1.43	−3.58	−2.44	−4.92	−1.00
Złotoria	4	−4.95	−3.07	−0.27	−1.80	0.05	−2.64	−1.94	−3.10	−0.52
Tykocin	5	−4.72	−2.69	0.52	−1.74	0.35	−3.92	−2.46	−3.95	−0.38
Wizna	6	−3.98	−2.35	−0.66	−2.03	−0.75	−3.33	−2.02	−3.69	0.15
Łomża	7	−4.08	−2.36	−0.93	−1.77	0.34	−2.83	−1.59	−3.47	0.53
Ostrołęka	8	−4.32	−3.22	−1.22	−2.09	−0.99	−3.83	−2.56	−4.85	−0.98
Różan	9	−2.97	−2.53	−0.78	−1.19	−0.14	−2.49	−1.78	−3.61	0.02
Pułtusk	10	−4.19	−2.98	−1.37	−1.88	−0.77	−3.92	−2.20	−5.11	−1.38
Serock	11	−2.78	−3.91	−1.98	−2.00	−2.11	−5.01	−2.93	−7.21	−5.69

Positive Igeo values for Cd were observed in sediments at the following points: Bondary (1.05), Wizna (0.15), Łomża (0.53), and Różan (0.02). A positive Igeo value for Pb was recorded at Tykocin, with a value of 0.52, classifying the sediment as class 1 (moderately polluted). For Zn, positive Igeo values were recorded at Bondary (0.08), Złotoria (0.05), Łomża (0.34), and Tykocin (0.35), also placing these sediments in class 1 (moderately polluted).

Most of the sediments listed were classified as class 1 (moderately polluted), except for the sediments from Bondary, which were classified as class 2 (strongly polluted). The arithmetic mean Igeo values for the studied metals followed the order: Zn > Pb > Cd > Cu > Mn > Co > Fe > Ni > Cr. The Igeo results indicate that Zn, Pb, and Cd are the metals most enriched in the Narew sediments, whereas Ni and Cr are the least enriched.

3.2. Metal Content of Macrophyte

In this study, all plants exhibited lower levels of Mn and Cu compared to phytotoxic thresholds [41,42], whereas higher levels were observed for Cd, Ni, Co, Pb, Cr, and Fe.

The accumulation of heavy metals in aquatic plants depends on various physicochemical factors, such as metal bioavailability in water, uptake rate through plant membranes, the stability of metals in biotic and abiotic environments, distribution across different plant tissues, and the potential to deposit metals in tissues [46].

Plants absorb metals like Fe, Mn, Cu, Zn, and Ni in small amounts, essential for their physiological processes. Notably, Co can stimulate plant growth. In contrast, metals such as Cd, Cr, and Pb have no known biological function and can be absorbed by plants, exhibiting toxicity even at deficient concentrations [47]. This variability underscores the influence of environmental and biological factors on aquatic plants' heavy metal uptake dynamics.

The Fe content in the analysed plants ranged from mg/kg, with an average concentration of 6178 mg/kg. *Potamogeton acutifolius* had the highest Fe content, while *Potamogeton nodosus* Poin contained the lowest amount of this element. The Mn content in the studied plants showed little variability, with an average concentration of 426 mg/kg. *Potamogeton acutifolius* (493 mg/kg) from the Korpacz site contained the highest Mn content, followed by *Potamogeton natans* L. (491 mg/kg) from the Wizna site. In contrast, the lowest Mn concentration was found in *Phragmites australis* (259 mg/kg) at the Bondary site. Zn concentrations ranged from 50 to 108 mg/kg, averaging 74 mg/kg. The highest Zn content was recorded in *Potamogeton nodosus* Poin. (107.59 mg/kg). Moreover, *Potamogeton natans* L. (104 mg/kg) at the Wizna site, followed by *Nuphar lutea* L. (103 mg/kg) at the Różan site. Cu content in plants ranged from 9.45 to 27.67 mg/kg, with *Potamogeton nodosus* Poin, showing the highest concentrations. Plants demonstrated good bioaccumulation capacity for Cu, with *Phragmites australis* (24.20 mg/kg) and *Phalaris arundinacea* L. (23.02 mg/kg) exhibiting high values and an average Cu concentration of 17.47 mg/kg for the studied species. The average Ni concentration was 6.15 mg/kg. *Potamogeton natans* L. (11.35 mg/kg) contained the highest amount of Ni, followed by *Potamogeton nodosus* Poin. (10.42 mg/kg). The lowest Ni concentration was observed in *Phalaris arundinacea* L., only 1.30 mg/kg. Co concentrations in macrophytes were less variable, with 77% values ranging from 1.86 to 6.51 mg/kg and an average of 3.22 mg/kg. Among the studied species, the highest Co concentrations were found in *Potamogeton perfoliatus* L. (6.51 mg/kg), *Potamogeton acruifolius* (5.12 mg/kg) from the Korpacz site, and *Potamogeton natans* L. (4.52 mg/kg). The highest Cr concentrations were recorded in *Potamogeton nodosus* Poin. (5.83 mg/kg) and *Potamogeton natans* L. (4.18 mg/kg) at the Wizna site, with an average Cr concentration of 2.54 mg/kg. The highest Pb concentrations were found in *Phragmites australis* (12.28 mg/kg) and *Potamogeton perfoliatus* L. (11.99 mg/kg), with an average of 8.37 mg/kg. Plant Cd concentrations averaged 0.22 mg/kg, ranging from 0.03 to 0.83 mg/kg. Among the studied species, *Potamogeton perfoliatus* L. (0.86 mg/kg) and *Potamogeton natans* L. (0.44 mg/kg) had the highest Cd concentrations. The study showed that *Potamogeton natans* L. (Ni 11.35, Cr 4.18, Zn 104, Cd 0.44, Co 4.52 mg/kg) is one of the analysed plant species that can accumulate large amounts of multiple trace elements. Of particular importance is the species Phragmites australis, which has been used in biological sewage treatment plants for several decades thanks to its high capacity to accumulate heavy metals such as Pb, Zn, and Cd. It is widely used in phytoremediation systems, emphasising its importance in purifying the aquatic environment [12,47].

The analysis of the research results identified the following sources of pollution in the study area. Plants from sediments primarily absorb Fe and Mn and are generally present in higher concentrations than other microelements [41]. Fe and Mn in sediments mainly originate from natural sources, such as the Earth's crust [48]. However, in ecosystems subject to anthropogenic pressure, such as the study area, the content of these elements may be influenced by additional sources, such as emissions from transport (road dust) and industrial activities. Fe is essential for aquatic plants' metabolic processes and proper functioning. Kabata-Pendias and Pendias [41] emphasise that Fe is characterised by low plant mobility, which may affect its distribution in various tissues. Aquatic plants can directly absorb iron through submerged leaves and stems, particularly in chelates (e.g., Fe-EDTA) or dissolved Fe^{2+} ions in water. This process is more efficient when iron is available in highly soluble forms and at optimal concentrations. Moreover, aquatic plants can modify the chemical conditions around their tissues by altering pH or releasing organic compounds to facilitate iron uptake. Some species can also produce siderophores—iron-chelating compounds—that enhance the solubility and bioavailability of iron in aquatic environments [9,46]. Mazmudar and Das [49] suggest that the high accumulation of Fe in aquatic plants is due to the reduction of insoluble Fe(III) to the more soluble Fe(II) in aquatic environments. The availability of Fe in sediments and water can promote its transport to the aerial parts of plants, particularly under conditions of increased bioavailability. One source of Fe pollution in rivers is transportation, as noted by Bardelli et al. [50], who highlight that Fe is present in road dust originating from the wear of vehicle brake pads.

Manganese is an essential micronutrient for aquatic plants, and its uptake occurs through multiple pathways. Bottom sediment Mn^{2+} ions, which are more bioavailable under anaerobic conditions often found in sediments, can be absorbed by the roots of rooted macrophytes, such as the common reed (Phragmites australis). In addition, Mn can be directly absorbed from water through submerged leaves and stems, particularly in the form of dissolved Mn^{2+} ions. This mechanism is significant for submerged plants. Mn is more accessible in water at low pH and under anaerobic conditions. Aquatic plants also exhibit adaptive mechanisms to enhance Mn uptake [9]. For example, they can reduce oxidised insoluble forms of Mn (Mn^{3+} or Mn^{4+}) to the more soluble Mn^{2+} by releasing reductive enzymes or organic substances, such as organic acids. The availability of Mn in aquatic environments depends on several factors, including redox conditions in sediments and water (with Mn^{2+} dominating in reducing conditions), water pH, and organic substances capable of complexing Mn [51]. Both Fe and Mn in rivers are primarily of natural origin. However, Mn pollution is also associated with anthropogenic activities such as transportation (tire wear and dust resuspension). Karar et al. [52] link Mn in soil and sediments to manganese cyclopentadienyl tricarbonyl (MMT). Zinc, being more mobile than Mn, plays a key role in the enzymatic activities in aquatic plants. It is transported from sediments and water in the form of Zn^{2+} ions, which can lead to excessive uptake by macrophytes, particularly in rivers near roads with high traffic density [53]. Zinc is emitted through tire wear during vehicle use [54] and may also enter surface waters via wastewater.

Copper in bottom sediments is mainly bound to organic matter and less mobile forms, such as sulphates, sulphides, and carbonates, which limits its accumulation in aquatic plants [55]. An analysis of copper content in macrophytes from the Narew River, which flows through urban areas, indicates that one of the sources of this element is road dust enriched in copper. Copper is also present in vehicle components, such as bearings, brake pads, engine parts, and radiators. The roots of plants rooted in bottom sediments play a significant role in copper uptake, mainly in the form of Cu^{2+} ions (oxidised form) or complex chelate compounds. Copper can be adsorbed from bottom sediments, especially in places with higher concentrations of this element or where organic compounds are present that increase its bioavailability. In anoxic conditions, such as deeper sediment layers, copper may be available in sulphide complexes (e.g., CuS), but the bioavailability of these forms is limited [51,56].

Submerged leaves and shoots can take up copper directly from water as dissolved Cu^{2+} ions or as chelates, such as complexes with humic or fulvic acids. Copper uptake by leaves and shoots is significant for plants floating on the water surface or wholly submerged, where root activity is limited or absent. Aquatic plants can also influence local chemical conditions to facilitate copper uptake; for example, by regulating the ambient pH, which increases copper solubility, or by secreting organic substances, such as organic acids, which improve the bioavailability of the element. Some plants accumulate copper in unique structures, which helps to minimise its toxic effects [21,27].

The uptake of copper by aquatic plants depends on many environmental factors, such as water pH because copper is more bioavailable in acidic environments, as well as the presence of organic compounds, the concentration of copper in water and sediments, and the redox potential, which affect the chemical form of Cu. These adaptive mechanisms allow aquatic plants to effectively take up and manage copper resources, even under variable environmental conditions [11,31].

Although Ni generally exhibits the lowest mobility and bioavailability among heavy metals at neutral to alkaline pH [21], its mobility can increase significantly under more acidic conditions [41]. Elevated Ni levels in aquatic plants from agricultural areas, including Poland [41], may result from compound fertilisers and pesticides [57,58]. Chromium is a widely distributed heavy metal in aquatic ecosystems, with emissions mainly stemming from anthropogenic sources. Cr levels in the environment depend on the proximity to pollution sources and meteorological factors, as it can travel considerable distances through the air [59]. Hexavalent Cr poses the most significant environmental threat due to its toxicity, while trivalent Cr is relatively less mobile [60–62]. Bhattacharya et al. [54] found significant amounts of Cr in road dust, originating from tire wear, road markings, and corrosion-resistant coatings on vehicles and barriers.

Lead (Pb) is relatively immobile in bottom sediments. Pb sources include transportation, as despite the use of unleaded fuel, Pb is still present in the environment. Rajaram et al. [63] indicate that Pb mainly enriches urban areas. In agricultural areas, higher Pb concentrations may result from using pesticides, phosphate fertilisers, and compound fertilisers [57,58]. Coal combustion can also contribute to increased Pb levels in aquatic ecosystems.

These findings underscore the importance of aquatic plants as indicators of ecosystem health and heavy metal pollution. Macrophytes, such as those studied in the Narew River, play a crucial role in metal accumulation, enabling the assessment of the impacts of both anthropogenic and natural pollutants on aquatic environments.

Cd, in contrast, exhibits high mobility [41] and is a dangerous metal. According to Zhou et al. [64], Cd causes various physiological and biochemical changes in plants. Sources of Cd include agriculture (pesticides, mineral and organic fertilisers, sewage sludge, manure, lime [65]) and transportation. Cd is released in dust from tire wear [60] and diesel engines [66]. The small particle size of Cd facilitates its long-distance transport.

Metals in natural aquatic ecosystems occur in low concentrations, depending on catchment geology, but human activities can significantly increase their levels. Industrial activities, fertilisation, industrial and municipal wastewater, mineral extraction, transportation, rainwater, and atmospheric deposition contribute to this issue. Consequently, this problem has become a focal point for research investigating the fate of metals in aquatic environments.

3.3. Identification of Pollution Sources Using Statistical Analyses

Cr, Cd, Cu, Zn, and Co concentrations in sediments are strongly correlated, with correlation coefficients ranging from r = 0.61 to r = 0.97. According to Ke et al. [43], Bhuyan et al. [44], and Ciazela et al. [45], consistent and robust correlations between metal pairs (r = 0.71 to r = 0.92) suggest a common origin. However, Pb is only correlated with Zn (r = 0.73). Most metals are correlated with Fe, with correlation strengths ranging from r = 0.63 to r = 0.83. Mn is correlated only with Cr (r = 0.68) and Fe (r = 0.75). Strong mutual correlations among metals indicate the shared sources of origin [67]. The correlations with Fe and Mn suggest the significant role of hydrated Fe and Mn oxides in binding metals in the studied sediments.

The lack of strong correlations between some metals suggests that their concentrations are not controlled by a single factor but by a combination of geochemical phases and mixed associations [68,69]. According to Robertson et al. [70], correlations between specific components can help identify their potential sources.

Two factors were identified in the Factor Analysis (FA) of metal concentrations in sediments (Table 4), explaining 54% and 17% of the variance, respectively. The first factor is strongly correlated with Cr (0.89), Zn (0.92), Cd (0.71), Pb (0.70), Fe (0.72), and Mn (0.77). The second factor correlates with Ni (0.87) and Cu (0.86). Poorly treated industrial and municipal wastewater and past transportation activities may have contributed to the enrichment of these metals, with phosphate fertilisers being a potential source of Cd. Hydrated Fe and Mn oxides play an active role in binding these metals in bottom sediments. Elevated Mn levels could also be associated with applying organic fertilisers to arable soils for horticultural and agricultural purposes [71].

Table 4. Classification of examined elements according to multivariate analysis (Factor Analysis—FA).

Variable	Sediments		Macrophytes	
	Factor 1	Factor 2	Factor 1	Factor 2
Ni	−0.26	0.87	0.08	0.56
Cr	0.89	0.31	0.89	0.07
Zn	0.92	0.12	0.71	−0.22
Cd	0.71	0.65	0.88	0.28
Co	0.61	0.66	0.86	0.35
Cu	0.32	0.86	0.59	0.53
Pb	0.70	−0.22	0.19	0.88
Fe	0.72	0.48	0.72	0.10
Mn	0.77	0.12	0.61	−0.68
Variance Explained [%]	54	17	47	21

As for the second factor and the associated loadings, these metals (Ni and Cu) likely originate from anthropogenic sources. In this context, plant protection products from agricultural areas could be the primary cause of Cu enrichment in bottom sediments.

The dynamics of factor value changes are presented in Figures 2 and 3, illustrating the variations of the identified factors across the studied locations. This tool is handy for representing the linear variability of detected factors correlated with specific metals. Notable locations for Factor 1 include the Bondary site, situated directly downstream of the Siemianówka Reservoir, which is presumed to be a source of metals. The Złotoria site is located downstream of the confluence of the Supraśl River with the Narew. The Supraśl River carries pollutants from Białystok, which explains the significant difference in factor values between the Suraż and Złotoria sites.

Another notable location is Łomża, the largest city along the Narew River. Łomża hosts a municipal wastewater treatment plant. Based on FAs and factor values, specific sites were identified as having the most significant contributions to the enrichment of bottom sediments with metals associated with Factor 1. Factor 2 accounts for only approximately 17% of the variability and has a minor influence compared to Factor 1.

The highest factor values representing the second factor, correlated with Cu and Ni, were recorded at the Bondary and Różan sites. Bondary is a location influenced by the Siemianówka Reservoir, whose impact was also noted in the discussion of the first factor. This area is surrounded by agricultural fields and traversed by a railway line. The Różan site is in a region dominated by grasslands where intensive agricultural practices are conducted. Approximately 1 km from the Narew River near Różan, the National Radioactive Waste Repository is located, which may further contribute to the observed metal concentrations.

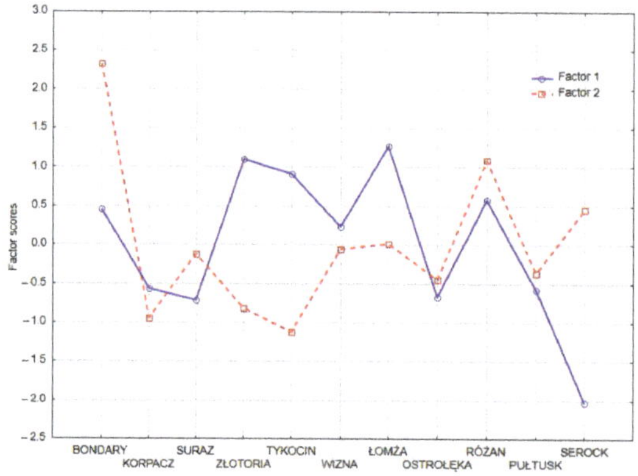

Figure 2. Factor scores in points—bottom sediments.

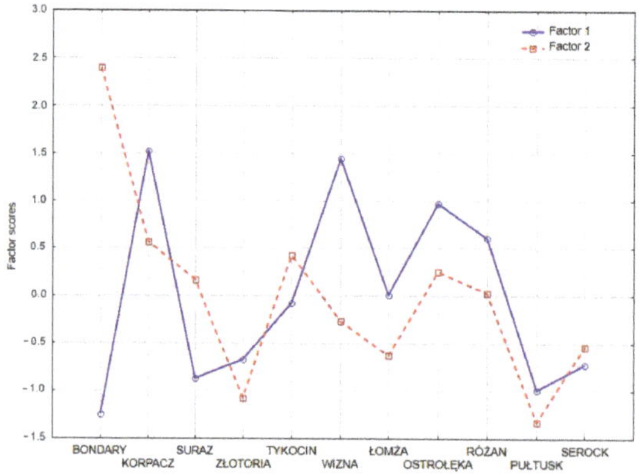

Figure 3. Factor scores in points—macrophytes.

Table 4 presents the FA results for metal concentrations in macrophytes. The number of factors and the distribution of factor loadings are similar to those obtained for bottom sediments. Factor 1 is correlated with Cr, Zn, Cd, Co, and Fe, explaining 46% of the variability, while Factor 2 is correlated with Pb, explaining 21%. A significant difference in description between sediments and macrophytes occurred for Factor 2. Factor 2 correlates with Pb, whose origin is likely associated with vehicular and railway traffic. A large portion of Pb is transported through the air, and macrophytes absorb the dry deposition of Pb via their leaves and stems. Aquatic plants selectively take up trace elements from their surrounding environment through their root system or leaves [72–74].

During the correlation analysis of metals in macrophytes, no correlations were observed between Pb, Ni, and Mn with other metals, confirmed by FA. This suggests that the origins of these elements differ. The factor value distribution for Factor 1 showed two maxima: Korpacz and Wizna. This may result from different migration pathways of metals to plants compared to sediments. In the case of the factor value distribution for Factor 2, a notable decrease in values was observed from the first point at Bondary, bordering the Siemianówka Reservoir, to the Złotoria point. Factor 2 is associated with Pb.

An objective Ward's cluster analysis was performed separately for sediments and plants. For sediments (Figure 4), the cluster analysis revealed two points where the bottom sediments had similar metal concentrations. The plant cluster analysis (Figure 5) also showed two groups of points with similar metal concentrations. The similarity of the point groupings for sediments and plants suggests a comparable response of sediments and macrophytes to metal source influences within the Narew River.

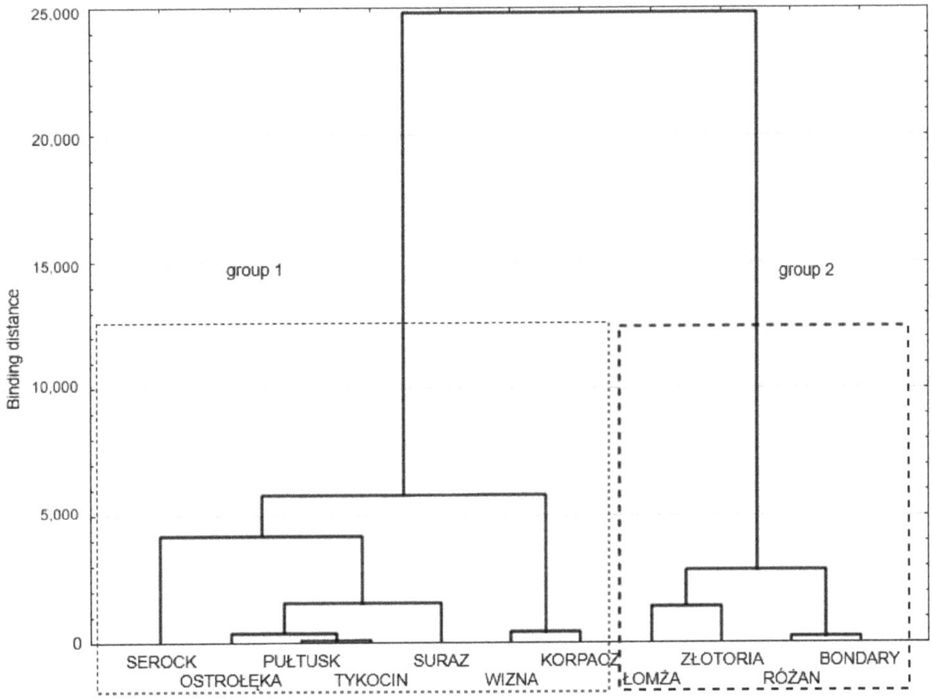

Figure 4. Hierarchical dendrograms for heavy metals in sediments obtained by Ward's hierarchical clustering method.

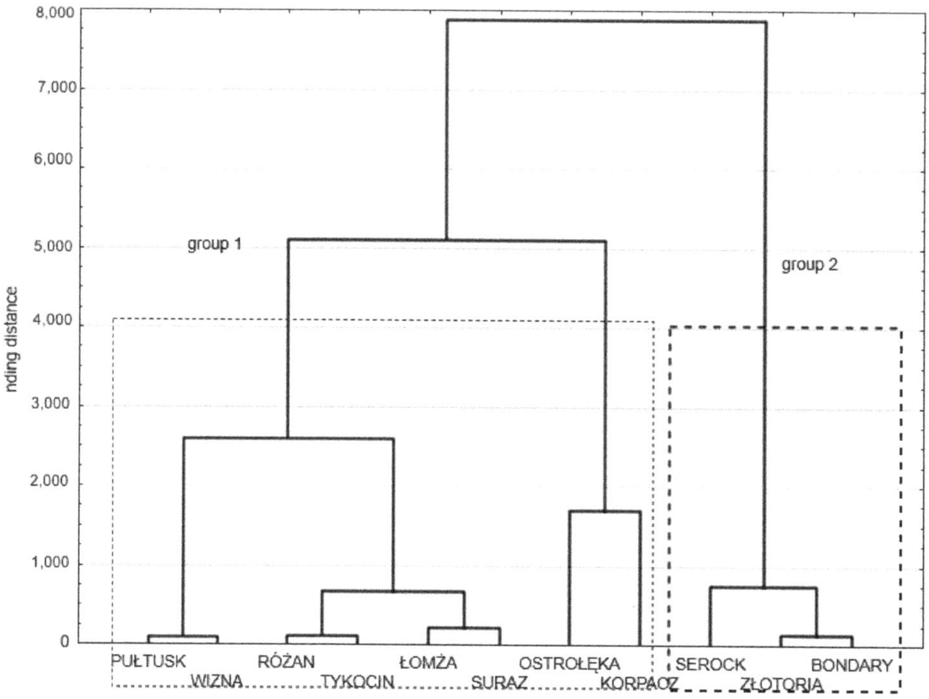

Figure 5. Hierarchical dendrograms for heavy metals in macrophytes were obtained by Ward's hierarchical clustering method.

4. Conclusions

The content of analysed heavy metals in bottom sediments and macrophytes of the Narew River is influenced by the watershed topography, hydrological conditions, and anthropogenic impacts. The average concentrations of heavy metals in bottom sediments decreased in the following order: Fe > Mn > Zn > Cu > Pb > Cr > Ni > Co > Cd. Analyses revealed that Ni, Cr, Co, Fe, and Mn concentrations remain at natural levels, whereas Zn, Cd, Cu, and Pb exhibit anthropogenic enrichment. The geochemical index indicated that most studied sediments belong to class 0 (uncontaminated). An exception was the Bondary sampling point, classified as class 2 (strongly contaminated), primarily due to high Cd concentrations.

The content of heavy metals in aquatic plants growing along the banks of the Narew River varied depending on the plant species and location. All examined species exhibited Mn and Cu levels below phytotoxic thresholds. In contrast, higher levels were recorded for Cd, Ni, Co, Pb, Cr, and Fe, indicating the diverse bioaccumulation capacities of different species. Mn and Cd in sediments and macrophytes demonstrated similar variability ranges, with low Mn variability indicating its natural origin and high Cd variability confirming its anthropogenic sources.

Correlation analysis showed that metals such as Cr, Cd, Cu, Zn, and Co are mutually correlated, suggesting their co-occurrence in the environment and shared sources of contamination. In contrast, Pb exhibited a distinct origin in macrophytes, emphasising its toxicity and a different distribution mechanism. The binding of metals in sediments by Fe and Mn compounds was confirmed by correlation coefficients (r = 0.63–0.83), underscoring their significant role in metal retention in the aquatic environment.

Ward's cluster analysis allowed for dividing sampling points into two characteristic groups for both bottom sediments and macrophytes, reflecting similar responses of these two environmental components to pollution sources. This study confirms that assessing the aquatic environment using both bottom sediments and macrophytes is more comprehensive and enables a more accurate estimation of pollution levels and identification of contamination sources. Monitoring heavy metals in bottom sediments and macrophytes is crucial for evaluating the overall condition of aquatic ecosystems. Integrating both approaches facilitates a better understanding of accumulation mechanisms, identification of anthropogenic pollution sources, and determination of environmental metal distribution patterns. The findings support the need for continued monitoring and the implementation of protective measures to preserve the high ecological value of the Narew River.

Author Contributions: Conceptualisation, M.S. (Mirosław Skorbiłowicz) and M.S. (Marcin Sidoruk); methodology, M.S. (Mirosław Skorbiłowicz); software, M.S. (Mirosław Skorbiłowicz) and M.S. (Marcin Sidoruk); validation, M.S. (Mirosław Skorbiłowicz) and M.S. (Marcin Sidoruk); formal analysis, M.S. (Mirosław Skorbiłowicz); investigation, M.S. (Marcin Sidoruk); writing—original draft preparation, M.S. (Mirosław Skorbiłowicz) and M.S. (Marcin Sidoruk); writing—review and editing, M.S. (Mirosław Skorbiłowicz) and M.S. (Marcin Sidoruk); visualisation, M.S. (Mirosław Skorbiłowicz) and M.S. (Marcin Sidoruk); supervision, M.S. (Mirosław Skorbiłowicz); project administration, M.S. (Mirosław Skorbiłowicz); funding acquisition, M.S. (Mirosław Skorbiłowicz) and M.S. (Marcin Sidoruk). All authors have read and agreed to the published version of the manuscript.

Funding: The research was carried out as part of research project No. WZ/WB-II 'S/2/2021 at Białystok University of Technology and financed from a subsidy provided by the Ministry of Science and Higher Education. The results presented in this paper were obtained as part of a comprehensive study financed by the University of Warmia and Mazury in Olsztyn, Faculty of Agriculture and Forestry, Department of Water Management and Climatology (grant no. 30.610.008–110). Funded by the Minister of Science under "the Regional Initiative of Excellence Pro- 584 gram References".

Data Availability Statement: The authors will make data available upon request.

Conflicts of Interest: The authors declare no conflicts of interest.

References

1. Brucet, S.; Poikane, S.; Lyche-Solheim, A.; Birk, S. Biological Assessment of European Lakes: Ecological Rationale and Human Impacts. *Freshw. Biol.* **2013**, *58*, 1106–1115. [CrossRef]
2. Ruiz-Hernandez, I.M.; Nouri, M.Z.; Kozuch, M.; Denslow, N.D.; Díaz-Gamboa, R.E.; Rodríguez-Canul, R.; Collí-Dulá, R.C. Trace element and lipidomic analysis of bottlenose dolphin blubber from the Yucatan coast: Lipid composition relationships. *Chemosphere* **2022**, *299*, 134353. [CrossRef]
3. Tlili, S.; Mouneyrac, C. New challenges of marine ecotoxicology in a global change context. *Mar. Pollut. Bull.* **2021**, *166*, 112242. [CrossRef]
4. Islam, M.S.; Ahmed, M.K.; Raknuzzaman, M.; Habibullah-Al-Mamun, M.; Islam, M.K. Heavy metal pollution in surface water and sediment: A preliminary assessment of an urban river in a developing country. *Ecol. Indic.* **2015**, *48*, 282–291. [CrossRef]
5. Abraham, J.S.; Sripoorna, S.; Choudhary, A.; Toteja, R.; Gupta, R.S.; Makhija, A.; Warren, A. Assessment of heavy metal toxicity in four species of freshwater ciliates (Spirotrichea: Ciliophora) from Delhi. *India. Curr. Sci.* **2017**, *113*, 2141. [CrossRef]
6. Zhang, M.; Chen, G.; Luo, Z.; Sun, X.; Xu, J. Spatial distribution, source identification, and risk assessment of heavy metals in seawater and sediments from Meishan Bay, Zhejiang coast. *China Mar. Pollut. Bull.* **2020**, *156*, 111217. [CrossRef]
7. Féraud, G.; Potot, C.; Fabretti, J.F.; Guglielmi, Y.; Fiquet, M.; Barci, V.; Maria, P.C. Trace elements as geochemical markers for surface waters and groundwaters of the Var River catchment (Alpes Maritimes, France). *Comptes Rendus Chim.* **2009**, *12*, 922–932. [CrossRef]
8. Moiseenko, T.I.; Gashkina, N.A. The Distribution of Trace Elements in Surface Continental Waters and the Character of Their Migration in Water. *Water Resour.* **2007**, *34*, 423–437. [CrossRef]
9. Sutkowska, K.; Teper, L.; Stania, M. Tracing potential soil contamination in the historical Solvay soda ash plant area, Jaworzno, Southern Poland. *Environ. Monit. Assess.* **2015**, *187*, 704. [CrossRef]

10. Ahmad, M.; Lee, S.S.; Lee, S.E.; Mohammad, I.; Al-Wabel, M.I.; Tsang, D.C.; Ok, Y.S. Biochar-induced changes in soil properties affected immobilisation/mobilisation of metals/metalloids in contaminated soils. *J. Soils Sediments* **2017**, *17*, 717–730. [CrossRef]
11. Ustaoğlu, F.; Kükrer, S.; Taş, B.; Topaldemir, H. Evaluation of metal accumulation in Terme River sediments using ecological indices and a bioindicator species. *Environ. Sci. Pollut. Res.* **2022**, *29*, 47399–47415. [CrossRef]
12. Vymazal, J.; Březinová, T. Accumulation of heavy metals in aboveground biomass of *Phragmites australis* in horizontal flow constructed wetlands for wastewater treatment: A review. *Chem. Eng. J.* **2016**, *290*, 232–242. [CrossRef]
13. Dai, L.; Wang, L.; Li, L.; Liang, T.; Zhang, Y.; Ma, C.; Xing, B. Multivariate Geostatistical Analysis and Source Identification of Heavy Metals in the Sediment of Poyang Lake in China. *Sci. Total Environ.* **2018**, *621*, 1433–1444. [CrossRef] [PubMed]
14. Shah, A.B.; Rai, U.N.; Singh, R.P. Correlations between Some Hazardous Inorganic Pollutants in the Gomti River and Their Accumulation in Selected Macrophytes under Aquatic Ecosystem. *Bull. Environ. Contam. Toxicol.* **2015**, *94*, 783–790. [CrossRef]
15. Harguinteguy, C.A.; Cofré, M.N.; Fernández-Cirelli, A.; Pignata, M.L. The Macrophytes *Potamogeton pusillus* L. and *Myriophyllum aquaticum* (Vell.) Verdc. as Potential Bioindicators of a River Contaminated by Heavy Metals. *Microchem. J.* **2016**, *124*, 228–234 [CrossRef]
16. Rezania, S.; Taib, S.M.; Din, M.F.M.; Dahalan, F.A.; Kamyab, H. Comprehensive Review on Phytotechnology: Heavy Metals Removal by Diverse Aquatic Plants Species from Wastewater. *J. Hazard. Mater.* **2016**, *318*, 587–599. [CrossRef] [PubMed]
17. Parzych, A.; Jonczak, J.; Sobisz, Z.; Woziwoda, B. The influence of plants on water quality in stream flowing through of mid-forest spring niche. *Desalin. Water Treat.* **2020**, *186*, 350–360. [CrossRef]
18. Bastviken, S.K.; Eriksson, P.G.; Premrov, A.; Tonderski, A. Potential denitrification in wetland sediments with different plant species detritus. *Ecol. Eng.* **2005**, *25*, 183–190. [CrossRef]
19. Duodu, G.O.; Goonetilleke, A.; Ayoko, G.A. Comparison of pollution indices for the assessment of heavy metal in Brisbane River sediment. *Environ. Pollut.* **2016**, *219*, 1077–1091. [CrossRef] [PubMed]
20. Milenkovic, N.; Damjanovic, M.; Ristić, M.D. Study of heavy metal pollution in sediments from the Iron Gate (Danube River), Serbia and Montenegro. *Pol. J. Environ. Stud.* **2005**, *14*, 781–787.
21. Kuang, Z.; Wang, H.; Han, B.; Rao, Y.; Gong, H.; Zhang, W.; Gu, Y.; Fan, Z.; Wang, S.; Huang, H. Coastal sediment heavy metal(loid) pollution under multifaceted anthropogenic stress: Insights based on geochemical baselines and source-related risks. *Chemosphere* **2023**, *339*, 139653. [CrossRef] [PubMed]
22. Calmuc, V.A.; Calmuc, M.; Arseni, M.; Topa, C.M.; Timofti, M.; Burada, A.; Iticescu, C.; Georgescu, L.P. Assessment of Heavy Metal Pollution Levels in Sediments and of Ecological Risk by Quality Indices, Applying a Case Study: The Lower Danube River, Romania. *Water* **2021**, *13*, 1801. [CrossRef]
23. Zonta, R.; Cassin, D.; Pini, R.; Dominik, J. Assessment of heavy metal and as contamination in the surface sediments of Po delta lagoons (Italy). *Estuar. Coast. Shelf Sci.* **2019**, *225*, 106235. [CrossRef]
24. Tomlinson, D.L.; Wilson, J.G.; Harris, C.R.; Jeffrey, D.W. Problems in the assessment of heavy-metal levels in estuaries and the formation of a pollution index. *Helgol. Mar. Res.* **1980**, *33*, 566. [CrossRef]
25. Ruiz, M.; Velasco, J. Nutrient Bioaccumulation in *Phragmites australis*: Management Tool for Reduction of Pollution in the Mar Menor. *Water Air Soil. Pollut.* **2010**, *205*, 173–185. [CrossRef]
26. Eid, E.M.; Shaltout, K.H.; Al-Sodany, Y.M.; Haroun, S.A.; Galal, T.M.; Ayed, H.; Khedher, K.M.; Jensen, K. Seasonal potential of *Phragmites australis* in nutrient removal to eliminate the eutrophication in Lake Burullus. *Egypt. J. Freshw. Ecol.* **2020**, *35*, 135–155. [CrossRef]
27. Du Laing, G.; Tack, F.M.G.; Verloo, M.G. Performance of selected destruction methods for the determination of heavy metals in reed plants (*Phragmites australis*). *Anal. Chim. Acta* **2003**, *497*, 191–198. [CrossRef]
28. Muller, G. Index of geoaccumulation in sediments of the Rhine River. *Geojournal* **1968**, *2*, 108–118.
29. Adimalla, N.; Wang, H. Distribution, contamination, and health risk assessment of heavy metals in surface soils from northern Telangana, India. *Arab. J. Geosci.* **2018**, *11*, 684. [CrossRef]
30. Kowalska, J.B.; Mazurek, R.; Gąsiorek, M.; Zaleski, T. Pollution indices as useful tools for the comprehensive evaluation of the degree of soil contamination—A review. *Environ. Geochem. Health* **2018**, *40*, 2395–2420. [CrossRef] [PubMed]
31. Pan, L.; Wang, Y.; Ma, J.; Hu, Y.; Su, B.; Fang, G.; Wang, L.; Xiang, B. A review of heavy metal pollution levels and health risk assessment of urban soils in Chinese cities. *Environ. Sci. Pollut. Res.* **2018**, *25*, 1055–1069. [CrossRef]
32. Zhaoyong, Z.; Mamat, A.; Simayi, Z. Pollution assessment and health risks evaluation of (metalloid) heavy metals in urban street dust of 58 cities in China. *Environ. Sci. Pollut. Res.* **2019**, *26*, 126–140. [CrossRef] [PubMed]
33. Sutkowska, K.; Czech, T.; Teper, L.; Krzykawski, T. Heavy metals soil contamination induced by historical zinc smelting in Jaworzno. *Ecol Chem. Eng. A.* **2013**, *20*, 1441–1450. [CrossRef]
34. Turekian, K.K.; Wedepohl, K.H. Distribution of the Elements in Some Mayor Units of the Earth's Crust. *Bull. Geol. Soc. Am.* **1961**, *72*, 175–184. [CrossRef]

35. Adimalla, N. Heavy metals pollution assessment and its associated human health risk evaluation of urban soils from Indian cities: A review. *Environ. Geochem. Health* **2020**, *42*, 173–190. [CrossRef]
36. Gopal, V.; Kalpana, G.; Nethaji, S.; Jayaprakash, M. Geochemical study of core sediments from Ennore Creek, North of Chennai, Tamil Nadu, India. *Arab. J. Geosci.* **2016**, *9*, 141. [CrossRef]
37. Lu, X.; Zhang, X.; Li, L.; Chen, H. Multivariate statistical analysis of heavy metals in street dust of Baoji, NW China. *J. Hazard. Mater.* **2010**, *173*, 744–749. [CrossRef]
38. Chen, T.; Liu, X.; Zhu, M.; Zhao, K.; Wu, J.; Xu, J.; Huang, P. Identification of trace element sources and associated risk assessment in vegetable soils of the urban-rural transitional area of Hangzhou, China. *Environ. Pollut.* **2011**, *159*, 1317–1323. [CrossRef] [PubMed]
39. Caeiro, S.; Costa, M.H.; Ramos, T.B.; Fernandes, F.; Silveira, N.; Coimbra, A.; Medeiros, G.; Painho, M. Assessing heavy metal contamination in Sado Estuary sediment: An index analysis approach. *Ecol. Indic.* **2005**, *5*, 151–169. [CrossRef]
40. Lee, C.S.; Li, X.D.; Shi, W.Z.; Cheung, S.C.; Thornton, I. Metal contamination in urban, suburban, and country park soils of Hong Kong: A study based on GIS and multivariate statistics. *Sci. Total Environ.* **2006**, *356*, 45–61. [CrossRef] [PubMed]
41. Kabata-Pendias, A.; Pendias, H. *Trace Elements in Soils and Plants*, 3rd ed.; CRC Press: Boca Raton, FL, USA, 2001.
42. Markert, B. *Plants as Biomonitors: Indicators for Heavy Metals in the Terrestrial Environment*; VCH Publishers: Weinheim, Germany, 1992.
43. Ke, X.; Gui, S.; Huang, H.; Zhang, H.; Wang, C.; Guo, W. Ecological risk assessment and source identification for metals in surface sediment from the Liaohe River protected area, China. *Chemosphere* **2017**, *175*, 473–481. [CrossRef] [PubMed]
44. Bhuyan, M.S.; Bakar, M.A.; Akhtar, A.; Hossain, M.B.; Ali, M.M.; Islam, M.S. Metal contamination in surface water and sediment of the Meghna River, Bangladesh. *Environ. Nanotechnol. Monit. Manag.* **2017**, *8*, 273–279. [CrossRef]
45. Ciazela, J.; Siepak, M.; Wojtowicz, P. Tracking metal contamination in a complex river-oxbow lake system: Middle Odra Valley, Germany/Poland. *Sci. Total Environ.* **2018**, *616–617*, 996–1006. [CrossRef] [PubMed]
46. Nazir, R.; Khan, M.; Masab, M.; Rehman, H.U.; Rauf, N.U.; Shahab, S.; Ameer, N.; Sajed, M.; Ullah, M.; Rafeeq, M. Accumulation of heavy metals (Ni, Cu, Cd, Cr, Pb, Zn, Fe) in the soil, water and plants and analysis of physico-chemical parameters of soil and water collected from Tanda Dam Kohat. *J. Pharm. Sci. Res.* **2015**, *7*, 89.
47. Peralta-Videa, J.R.; Lopez, M.L.; Narayan, M.; Saupe, G.; Gardea-Torresdey, J. The biochemistry of environmental heavy metal uptake by plants: Implications for the food chain. *Int. J. Biochem. Cell Biol.* **2009**, *41*, 1665–1677. [CrossRef] [PubMed]
48. Shabanov, M.V.; Marichev, M.S.; Minkina, T.M.; Rajput, V.D.; Bauer, T.V. Change in eco-geochemistry of bottom sediments in copper–pyrite mining area: A case-study of Karabash copper deposits. *Mining Inf. Anal. Bull.* **2023**, *5-1*, 117–134.
49. Mazmudar, K.; Das, S. Phytoremediation of Pb, Zn, Fe, and Mg with 25 wetland species from a papermill contaminated site in North East India. *Environ. Sci. Pollut. Res.* **2015**, *22*, 701–710. [CrossRef]
50. Bardelli, F.; Cattaruzza, E.; Gonella, F.; Rampazzo, G.; Valotto, G. Characterization of road dust collected in Traforo del San Bernardo highway tunnel: Fe and Mn speciation. *Atmos. Environ.* **2011**, *45*, 6459–6468. [CrossRef]
51. Baldantoni, D.; Alfani, A.; Di Tommasi, P.; Bartoli, G.; Virzo De Santo, A. Assessment of macro and microelement accumulation capability of two aquatic plants. *Environ. Pollut.* **2004**, *130*, 149–156. [CrossRef]
52. Karar, K.; Gupta, A.K.; Kumar, A.; Biswas, A.K. Characterization and identification of the sources of chromium, zinc, lead, cadmium, nickel, manganese and Iron in PM10 particulates at the two sites of Kolkata, India. *Environ. Monit. Assess.* **2006**, *120*, 347–360. [CrossRef] [PubMed]
53. Minkina, T.; Fedorenko, A.; Nevidomskaya, D.; Fedorenko, G.; Pol'shina, T.; Sushkova, S.; Chaplygin, V.; Beschetnikov, V.; Dudnikova, T.; Chernikova, N.; et al. Uptake of potentially toxic elements and polycyclic aromatic hydrocarbons from the hydromorphic soil and their cellular effects on the Phragmites australis. *Environ. Pollut.* **2023**, *309*, 119727. [CrossRef]
54. Bhattacharya, T.; Chakraborty, S.; Tuteja, D.; Patel, M. Zinc and chromium load in road dust, suspended particulate matter and foliar dust deposits of Anand City, Gujarat. *Open J. Met.* **2013**, *3*, 42–50. [CrossRef]
55. Terelak, H.; Motowicka-Terelak, T.; Stuczyński, T.; Pietruch, C. *Pierwiastki Śladowe (Cd, Cu, Ni, Pb, Zn) W Glebach Użytków Rolnych Polski [Trace Elements (Cd, Cu, Ni, Pb, Zn) in Agricultural Soils of Poland]*; IUNG: Pulawy, Poland, 2000.
56. Adamiec, E. Road environments: Impact of metals on human health in heavily congested cities of Poland. *Int. Environ. Res. Public Health* **2017**, *14*, 697. [CrossRef]
57. Kelepertzis, E. Accumulation of heavy metals in agricultural soils of Mediterranean: Insights from Argolida basin, Peloponnese, Greece. *Geoderma* **2014**, *221–222*, 82–90. [CrossRef]
58. Srivastava, V.; Sarkar, A.; Singh, S.; Singh, P.; de Araujo, A.S.F.; Singh, R.P. Agroecological responses of heavy metal pollution with special emphasis on soil health and plant performances. *Front. Environ. Sci.* **2017**, *5*, 64. [CrossRef]
59. Perugini, M.; Manera, M.; Grotta, L.; Abete, M.C.; Tarasco, R.; Amorena, M. Heavy metal (Hg, Cr, Cd, and Pb) contamination in urban areas and wildlife reserves: Honeybees as bioindicators. *Biol. Trace Elem. Res.* **2011**, *140*, 170–176. [CrossRef]

60. Gope, M.; Masto, R.E.; George, J.; Hoque, R.R.; Balachandran, S. Bioavailability and health risk of some potentially toxic elements (Cd, Cu, Pb and Zn) in street dust of Asansol, India. *Ecotoxicol. Environ. Saf.* **2017**, *138*, 231–241. [CrossRef] [PubMed]
61. Siedlecka, A.; Tukendorf, A.; Skórzyńska-Polit, E.; Maksymiec, W.; Wójcik, M.; Baszyński, T.; Krupa, Z. Angiosperms (Asteraceae, Convolvulaceae, Fabaceae and Poaceae; other than Brassicaceae). In *Metals in the Environment*; Analysis by Biodiversity; Prasad, M.N.V., Ed.; Marcel Dekker, Inc.: New York, NY, USA, 2001; pp. 171–217.
62. Bonanno, G. Comparative performance if trace element bioaccumulation and biomonitoring in the plant species Typha domingensis, *Phragmites australis* and *Arundo donax*. *Ecotoxicol. Environ. Saf.* **2013**, *97*, 124–130. [CrossRef] [PubMed]
63. Rajaram, B.S.; Suryawanshi, P.V.; Bhanarkar, A.D.; Rao, C.V.C. Heavy metals contamination in road dust in Delhi city, India. *Environ. Earth Sci.* **2014**, *72*, 3929–3938. [CrossRef]
64. Zhou, C.; Ge, N.; Guo, J.; Zhu, L.; Ma, Z.; Cheng, S.; Wang, J. *Enterobacter asburiae* Reduces Cadmium Toxicity in Maize Plants by Repressing Iron Uptake-Associated Pathways. *J. Agric. Food Chem.* **2019**, *67*, 10126–10136. [CrossRef]
65. Niassy, S.; Diarra, K. Effects of Organic Inputs in Urban Agriculture and Their Optimisation for Poverty Alleviation in Senegal, West-Africa. In *Organic Fertilizers: Types, Production and Environmental Impact*; Singh, R.P., Ed.; Nova Science Publishers: New York, NY, USA, 2012; pp. 1–22.
66. Nagajyoti, P.C.; Lee, K.D.; Sreekanth, T.V.M. Heavy metals, occurrence and toxicity for plants: A review. *Environ. Chem. Lett.* **2010**, *8*, 199–216. [CrossRef]
67. Azhar, M.; Muhammad, Z.C.; Nasir, A.; Tariq, J.; Abdul, G. Metal pollution and ecological risk assessment in marine sediments of Karachi Coast, Pakistan. *Environ. Monit. Assess.* **2013**, *185*, 1555–1565.
68. Jain, C.K.; Singhal, D.C.; Sharma, M.K. Metal pollution assessment of sediment and water in the river Hindon, India. *Environ. Monit. Assess.* **2005**, *105*, 193–207. [CrossRef] [PubMed]
69. Liu, C.B.; Xu, J.; Liu, C.G.; Zhang, P.; Dai, M.X. Heavy metals in the surface sediments in Lanzhou reach of Yellow River, China. *Bull. Environ. Contam. Toxicol.* **2009**, *82*, 26–30. [CrossRef] [PubMed]
70. Robertson, D.J.; Taylor, K.G.; Hoon, S.R. Geochemical and mineral magnetic characterisation of urban sediment particulates, Manchester, UK. *Appl. Geochem.* **2003**, *18*, 269–282. [CrossRef]
71. Sager, M. Urban soils and road dust-civilisation effects and metal pollution—A review. *Environments* **2020**, *7*, 98. [CrossRef]
72. Sawidis, T.; Chettri, M.K.; Papaionnou, A.; Zachariadis, G.; Stratis, J. A study of metal distribution from lignite fuels using trees as biological monitor. *Ecotoxicol. Environ. Saf.* **2001**, *48*, 27–35. [CrossRef] [PubMed]
73. Konstantinova, E.; Minkina, T.; Nevidomskaya, D.; Bauer, T.; Zamulina, I.; Latsynnik, E.; Dudnikova, T.; Yadav, R.K.; Burachevskaya, M.; Mandzhieva, S. Pollution and Ecological Risk Assessment of Potentially Toxic Elements in Sediments Along the Fluvial-to-Marine Transition Zone of the Don River. *Water* **2024**, *16*, 3200. [CrossRef]
74. Konstantinova, E.; Minkina, T.; Mandzhieva, S.; Nevidomskaya, D.; Bauer, T.; Zamulina, I.; Sushkova, S.; Lychagin, M.; Rajput, V.D.; Wong, M.H. Ecological and Human Health Risks of Metal–PAH Combined Pollution in Riverine and Coastal Soils of Southern Russia. *Water* **2023**, *15*, 234. [CrossRef]

water

MDPI

Article

Pollution and Ecological Risk Assessment of Potentially Toxic Elements in Sediments Along the Fluvial-to-Marine Transition Zone of the Don River

Elizaveta Konstantinova [1,*], Tatiana Minkina [1], Dina Nevidomskaya [1], Tatiana Bauer [1], Inna Zamulina [1], Elizaveta Latsynnik [1], Tamara Dudnikova [1], Rajendra Kumar Yadav [2], Marina Burachevskaya [1] and Saglara Mandzhieva [1]

1 Academy of Biology and Biotechnologies, Southern Federal University, Rostov-on-Don 344090, Russia; tminkina@mail.ru (T.M.); dnevidomskaya@mail.ru (D.N.); bauer@sfedu.ru (T.B.); inir82@mail.ru (I.Z.); lacynnik@sfedu.ru (E.L.); tyto98@yandex.ru (T.D.); marina.0911@mail.ru (M.B.); msaglara@mail.ru (S.M.)

2 Department of Soil Science, Agriculture University, Kota 324001, India; rajendra@aukota.org

* Correspondence: konstantliza@gmail.com

Abstract: The quality of sediments in the mixing zone of river freshwater and marine saline water as an important geochemical barrier for potentially toxic elements (PTEs) remains poorly understood. This study aims to analyze the current pollution with PTEs and associated ecological risks in sediments of the Don River delta and the surrounding area of the Taganrog Bay of the Sea of Azov (Russia). The PTE content was determined in fifty-four collected samples using the WDXRF and assessed using geochemical and ecotoxicological indicators. The source of Cr, Mn, Ni and Pb is mainly river runoff, and Cu, Zn and Cd are from a variety of anthropogenic sources. As shown by the assessment of the geoaccumulation index (*Igeo*), single pollution index (*PI*) and contamination factor (*CF*), these elements are the priority pollutants. According to these estimates, high and very high contamination of sediments in the estuarine zone of the Don River with Cd and Pb was detected in 72–94% and 2–57% of samples, respectively. However, environmental risks are determined almost exclusively by the level of Cd. Total contamination as assessed by the Nemerow pollution index (*NPI*), modified degree of contamination (mC_d) and metal pollution index (*MPI*) is of concern in 83–98% of the samples studied. The most heavily polluted sediments are in the vicinity of residential areas of the Taganrog Bay. Despite the lower average pollution levels of deltaic sediments, freshwater biota are exposed to higher potential toxic risks of adverse effects by PTE, particularly from Ni and Pb. Thus, the complex hydrological regime and uneven anthropogenic impact predetermine the geochemical state of the sediments of the estuarine zone of the Don River.

Keywords: heavy metals; trace elements; PTEs; surface sediments; sediment pollution; sediment contamination; pollution level; ecological risk; risk assessment; Southern Russia

1. Introduction

Over the past decades, there has been growing interest among researchers in studying the state of sediments of river-to-marine transition zones of large rivers based on the content of priority pollutants. The aquatic environments of large river deltas and estuarine zones significantly moderate the continual geochemical cycles, as well as control the transport of dissolved and suspended matter from vast watersheds to seas and oceans [1–5]. Despite the high importance of fluvial–marine systems for sustainable development and pollution mitigation at both the regional and global levels, a comprehensive analysis of their current status is quite difficult to achieve [6]. The following situation is predetermined by the dynamic nature of delta and coastal landscapes, as well as constant transformation due to the natural and anthropogenic changes in natural environments [6–8].

River freshwater infilling, waves and tides, coastal abrasion and storms in the sea–river transition zone, as well as erosional processes in watersheds, strongly affect the morphology and geochemistry of large river deltas and estuarine areas [9–12]. The above processes have a significant impact on aquatic and floodplain ecosystems, paddy soils and, most dramatically, sedimentation rate and chemical composition of riverbeds and nearshore sediments [13–18]. The latter is of particular interest since the geochemistry of estuarine sediments is important both for the assessment of historic erosion and pollution within a catchment [19] and for determining the contribution of specific sources and activities [12,20,21]. Another issue related to this component of the aquatic environment is the potential ability of sediments accumulated within delta channels, as well as coastal and near-shore zones of the marine part of the system, to become a source of secondary pollution [22]. The large amounts of pollutants, accumulated within the sediments of large river deltas in highly industrialized and populated regions of the world, can be subjected to remobilization due to strong storms [23], tsunamis [24,25] or other catastrophic natural or anthropogenically induced events [26], that could occur either upstream or at sea.

Analysis of the levels and potential environmental risks of most common pollutants in river delta sediments and adjacent marine areas is important for the sustainable management of these landscapes. Many recent studies focus on emerging pollutants, particularly polycyclic aromatic hydrocarbons [27–29], polychlorinated biphenyls [30,31], microplastics [32–34], etc. due to their high toxicity [35]. At the same time «traditional» pollutants such as potentially toxic elements (PTEs) can be of high concern and significantly contribute to the total pollution; additionally, their accumulation period duration can be very long, and may correspond to the preindustrial and early industrial periods [19,36].

The assessment of sediment pollution by PTEs and the associated potential risks of their negative impact on aquatic ecosystems involves several fundamental approaches. The first approach consists of directly assessing contamination as an excess level of PTE due to anthropogenic activity relative to the natural abundance of chemical elements in sediments, or the geochemical background [1,17,20,37,38]. The global background, in particular the content of elements in the upper continental crust (UCC) [17,36,39], sedimentary rocks, soils, the regional geochemical background corresponding to threshold or baseline values in pristine sediments formed under similar conditions [17,40,41], or the preindustrial reference value, are used as the normal concentration of PTEs. Sediment contamination with an individual PTE is determined by calculating various indicators, in particular the enrichment factor (*EF*), single pollution index (*PI*), geoaccumulation index (*Igeo*), contamination factor (*CF*), etc., characterizing the excess of their contents over the background, including the use of fractional or elemental normalization. Overall pollution is defined as a sum, arithmetic mean, geometric mean, or other metric assessed on a given scale [1,37,38]. This approach does not take into account the toxicity of PTEs and their fate in landscapes, but this drawback is overcome by assessing potential ecological risks using the Hakanson scheme [42]. Direct negative effects of pollutants from sediments on aquatic biota are assessed by biotesting, but this method is expensive and time-consuming. As an alternative, current levels of PTEs in sediments are compared using sediment quality guidelines (SQGs) [2], which are empirically derived from sediment chemistry, toxicity, bioaccumulation, and community structure data [35,43]. SQGs are those levels of pollutants at which, if exceeded, significant negative biological effects occur.

Recent geochemical studies of the ecosystems of the transition zone of the Don River–Taganrog Bay system demonstrate high levels of anthropogenic PTE in floodplain and coastal soils [44–47], vegetation [47–49], waters and suspended matter [50–54], marine bottom sediments [55–57] and freshwater sediments [31,58]. At the same time, an assessment of the quality of surface and bottom sediments of the coast of the Taganrog Bay and the Don Delta in terms of the level of PTEs and possible risks to the environment has not been carried out previously. In this context, this work aims to contribute to the study of geochemistry and the quality of sediments along the fluvial to marine transition zone of the Don River. To this end, physicochemical sediment parameters, levels and sources of

PTE and related pollution status, and ecological and ecotoxicological risks were analyzed and compared in sediments of the Don River delta and adjacent coast of the Taganrog Bay.

2. Materials and Methods

2.1. Study Area

Taganrog Bay is the largest in the shallow shelf inland Sea of Azov, located in Eastern Europe (Figure 1A). The part of the bay that juts out onto the shore contains the mouth of the largest river flowing into the Sea of Azov—the Don [4]. The modern geomorphology and sediments of the Azov region were formed in the Holocene during the last transgression of the Black Sea under conditions of uneven sea level rise with downward tectonic movements [59]. The region has a humid continental climate (Dfa according to the Köppen classification) with cold winters and dry, hot summers. North and north-eastern winds prevail during cold seasons and west to south-western during the warm season [60].

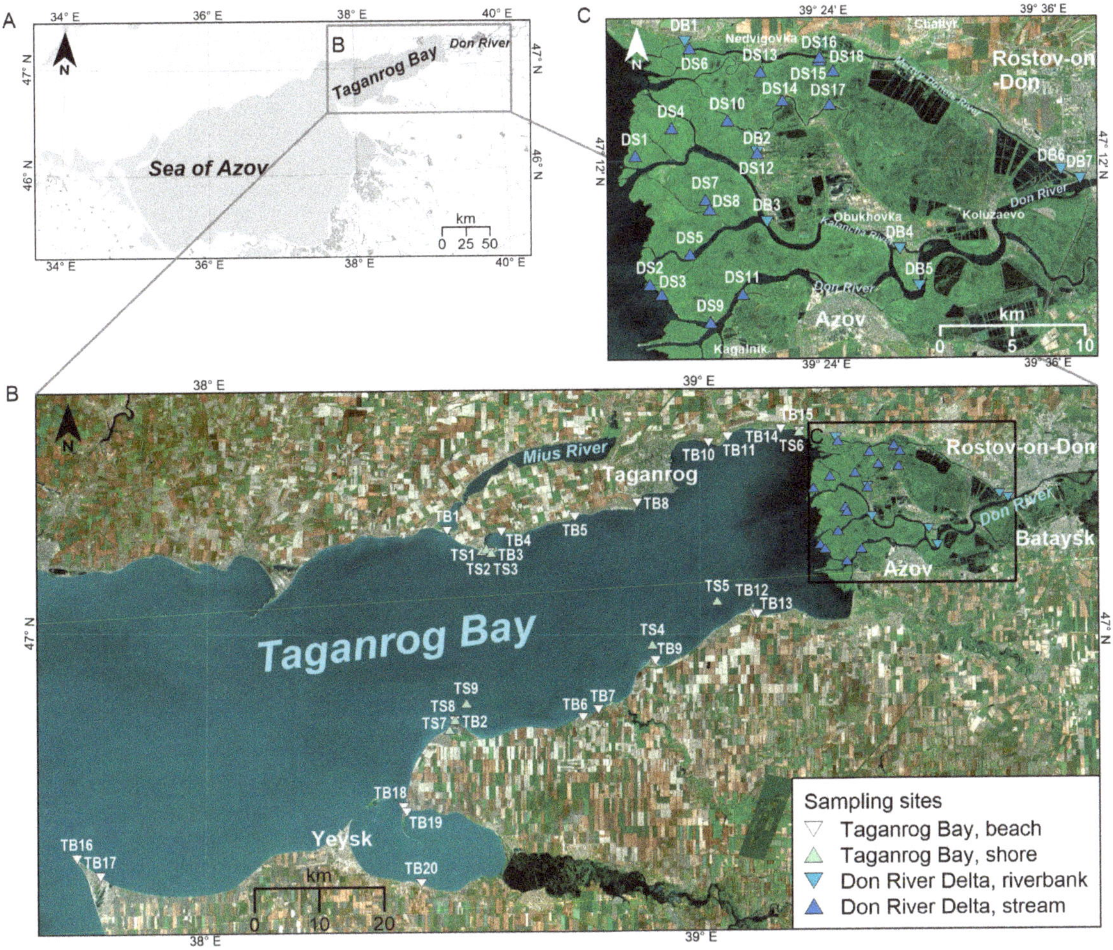

Figure 1. Location of sampling sites for surface and bottom sediments within the Sea of Azov basin (**A**), along the coast of the Taganrog Bay (**B**) and the Don River delta (**C**).

The study covers the Don River estuary area, including the Don River delta (Figure 1C) and the coast of the estuary from the marine edge of the delta to the marine boundary of the estuary area on the southern coast of the Taganrog Bay and from the edge of the delta to the mouth of the Mius River on the northern coast of the bay (Figure 1B).

The Taganrog Bay has an area of 5600 km^2 and an average depth of 4.9 m. The shores of the Taganrog Bay are composed mainly of loess-like loams and Scythian clays, which, due to abrasion, are the source of material for the formation of marine sediments. Depending on the prevailing geological processes, abrasion, abrasion-landslide and accumulative shores are formed [61]. The average annual sedimentation rate is 700 g m^{-2} per year [56]. The Taganrog Bay basin belongs to the Black Sea steppe province, where the zonal vegetation is represented by forb–fescue–feather grass steppes reaching the supralittoral. The sandy areas of the coast are deserts with a mosaic combination of halophyte communities, and meadows, aquatic coastal and floodplain vegetation develop in the river mouths [59]. The salinity of the waters of the bay varies from 0 to 14‰ [62]. The salinity gradient of the surface layer of water in the Taganrog Bay is on average 0.5–7‰ from the edge of the Don delta to the open part in the south [63].

The modern delta of the Don River occupies the coastal part of the flat alluvial delta plain from the branch of the main channel of the Don River, the Mertvy Donets River, to the sea edge of the land delta and occupies an area of 538 km^2 [64]. The delta floodplain is represented by low-lying islands (0.2–1 m a.s.l.) composed of medium- and fine-grained sands, separated by numerous branches and channels. The Don Delta has a wave-dominated regime and is subject to strong sea surges and erosion of the sea edge due to the construction of the Tsimlyansk Reservoir in 1953 (309 km from the mouth) [16]. Average annual water runoff and fluvial sediment delivery of the Don River are estimated at 20.6 km^3 and 0.19 Mt per year, respectively [16,60]. The Don Delta is characterized by fluctuations in water salinity depending on wind-driven phenomena [29]. Under normal conditions, fresh waters predominate (0.6–0.95‰); with strong westerly winds, water salinity can reach 5–6‰ [63]. The delta combines protected, natural, rural and urban landscapes. Most of the territory is subject to continuous overgrowth by monodominant communities of southern reed *Phragmites australis* Cav. Cattail groups of *Typha australis* Schum. and Thonn. and *Typha laxmannii* Lepech. are common on the marine edge of the delta [48,49].

Anthropogenic activity is a key factor in the deterioration of the quality of ecosystems in the transit estuarine zone of the Don River. High urbanization leads to the entry of pollutants into the waters of the Lower Don and Taganrog Bay with poorly treated wastewater and storm water from the cities of Volgodonsk, Novocherkassk, Aksay, Rostov-on-Don, Bataysk, Azov, Taganrog, Yeysk, etc. [31]. Industrial production is concentrated in the cities, including ferrous metallurgy, mechanical engineering and the chemical industry, the emissions of which contain hazardous substances [55]. Port infrastructure and maritime transportation are also an important source of water and sediment pollution [45,46,57]. The coastal areas of Taganrog Bay are occupied by agricultural lands, where erosion and surface runoff lead to the entry of fertilizers, pesticides, and manure runoff into marine ecosystems [59].

2.2. Sediment Sampling and Preparation

The sampling campaign was conducted from May to July 2023. Surface and bottom sediments in the Don River delta were collected in the floodplain on the river banks at a distance of 2–3 m from the water's edge (DB; $n = 7$) and in the channels of the Don River and its large tributaries (DS; $n = 18$), respectively. Surface and bottom sediments on the coast of Taganrog Bay were collected from beaches in the surf zone (TB; $n = 20$) and within the underwater coastal slope (TS; $n = 9$), respectively (Figure 1B,C, Table A1). Surface sediments were collected to a depth of 0–20 cm from five surface subsamples using a hand auger set (Eijkelkamp Agrisearch Equipment BV, Giesbeek, The Netherlands). Bottom sediments (approximately 10–15 cm thick) were collected using Van Veen modified grab sampler (Ocean-0.025, Remstroymash, Murmansk, Russia) with a surface area of 0.025 m^2 in triplicate from each sampling site. All samples were placed in clean polyethylene bag using a stainless-steel spatula. Collected samples were stored at 4 °C prior to further analysis. In the laboratory, the samples were dried in an oven at a temperature < 60 °C to prevent organic matter and chemical element loss, and were mixed, ground and passed through a 1 mm sieve [65].

2.3. Analytical Methods

Measurements of pH and electrical conductivity (EC) of the sediments were made in situ using a portable multimeter (WTW, Xylem Analytics, Weilheim, Germany).

Physicochemical properties were determined in all sediment samples: total organic carbon (TOC) by the modified Tyurin method [66], the $CaCO_3$ content by Kudrin's complexometric method, with the destruction of calcium carbonate with a 0.02 mol/L HCl solution and subsequent titration of excess acid with alkali [65], and sum of exchangeable cations Ca^{2+} and Mg^{2+} in a 1 N KCl extract determined by the Shaimukhametov method [67]. The fine clay fraction of sediments (<1 μm) was obtained using the sedimentation method proposed by Gorbunov [68].

The total concentrations of Al, Cr, Mn, Ni, Cu, Zn, Cd and Pb were determined using wavelength dispersive X-ray fluorescence (WDXRF) (Spectroscan MAX-GV, Spectron NPO Ltd., Saint Petersburg, Russia). Prior to WDXRF analysis, air-dried bulk samples were pulverized manually with an agate mortar and pressed into a tablet on a boric acid substrate. Quantitative analysis of chemical elements was carried out using Spektr-Kvant software 4 (Spectron NPO Ltd., Saint Petersburg, Russia). Detection limits were 10 mg/kg for Cr, Mn, Ni and Pb, 5 mg/kg for Cu and Zn, and 0.1 mg/kg for Cd. All laboratory tests were performed in triplicate. For quality assurance and quality control (QA/QC), the duplicates, reagent blanks, internal standards, and certified reference material GSS 3485-86 (Vinogradov Institute of Geochemistry SB RAS, Irkutsk, Russia) were used. Accuracy and precision corresponded to the standards of the certified method [69]. All the geochemical data obtained in this study are available in Table A2.

2.4. Pollution and Ecological Risk Assessment

2.4.1. Enrichment Factor

The enrichment factor (*EF*) is used as a means of assessing the possible origin of elements in soils [2,45], marine and river sediments [20,21,37,38,40,70–73], and suspended matter [37], and is calculated using Equation (1):

$$EF = (C_i/C_{Al})_{Sample}/(C_i/C_{Al})_{Background}, \quad (1)$$

where C_i and C_{Al} are the contents of a given element and Al, respectively. The background values were referenced to upper continental crust (UCC) [39] (Table A3). *EF* values > 10 indicate anthropogenic origins of PTEs in sediments [21]. According to *EF* values, enrichment can be divided into seven classes (Table A4).

2.4.2. Geoaccumulation Index

The geoaccumulation index (*Igeo*), introduced by Müller [74], is widely used to assess river and marine sediment pollution by individual inorganic pollutants [1,20,37,38,40,41,43,71,72]. *Igeo* is a measure of the level of contamination of sediments by pollutants relative to the standard, which is the geochemical background in clay rocks, taking into account the elimination of natural fluctuations in background levels [74] and is calculated according to (2):

$$Igeo = \log_2(C_i/1.5C_b), \quad (2)$$

where C_b is the regional geochemical background. The average PTE content in uncultivated *Haplic Chernozems* of the Northern Azov region were used as background data [48] (Table A3). Seven pollution classes are distinguished according to *Igeo* (Table A4).

2.4.3. Single Pollution Index and Nemerow Pollution Index

The single pollution index (*PI*) is the ratio of the concentration of an element in sediments to the natural background. The calculation of *PI* is the basic method for assessing the pollution of soils and sediments by individual elements and is used to calculate many complex indicators, in particular the Nemerow pollution index (*NPI*) [41,45,70,75]. The *NPI* takes into account the combined effect of the most hazardous pollutant along with the average total contamination. Both indices are calculated as follows (3) and (4):

$$PI = C_i/C_b, \tag{3}$$

$$NPI = \sqrt{\left(PI_{avg}^2 + PI_{max}^2\right)/2}, \tag{4}$$

where PI_{avg} and PI_{max} are the average and maximum *PI* values for a range of PTEs under consideration, respectively. The *PI* and *NPI* define one of five classes of individual and total sediment pollution (Table A4).

2.4.4. Contamination Factor and Modified Degree of Contamination

The contamination factor (*CF*) was presented by Hakansson [42] as a measure of the level of pollution of sediments, determined by comparing pollutant contents in modern surface sediments with pre-industrial levels in deeper sediment layers. The calculation is expressed as follows (5):

$$CF = C_i/C_0, \tag{5}$$

where C_0 is the 'pre-industrial' background. The average contents of PTEs in the 25–45 cm layer of bottom sediment columns of the Taganrog Bay were used as the pristine reference levels [50]. Since there are no data on total Zn and Cd content in the columns, its concentrations in the bottom sediments of the Taganrog Bay from 1993 were used [57] (Table A3). Four classes of pollution are distinguished according to the *CF* (Table A4).

Total pollution was estimated by Hakanson as the sum of *CF* for 8 substances and expressed as the degree of contamination (C_d) [42]. Although this assessment scheme was proposed for assessing the condition of lake sediments, it has now been successfully applied to marine and river sediments as well [20,37,38,41,72]. To take into account the overall contamination of sediments regardless of the number of substances assessed, Abrahim [71] proposed the modified degree of contamination (mC_d) corresponding to the arithmetic mean *CF* of all substances analyzed (6):

$$mC_d = \left(\sum_{i=1}^{n} CF_i\right)/n, \tag{6}$$

where '*i*' denotes the *ith* element and '*n*' represents the number of elements analyzed. The classification of *CF* and mC_d is presented in Table A4.

2.4.5. Metal Pollution Index

The metal pollution index (*MPI*) is an alternative to mC_d for assessing polyelement pollution of both abiotic environments [38,40,72] and living organisms [76]. The *MPI* is the geometric mean of the *CF* values of all analyzed metals (7):

$$MPI = \left(\prod_{i=1}^{n} CF_i\right)^{\frac{1}{n}} = \sqrt[n]{(CF_1 \times CF_2 \times \ldots \times CF_n)}, \tag{7}$$

where '*i*' denotes the *ith* element and '*n*' represents the number of elements analyzed. Three pollution classes are distinguished according to *MPI* (Table A4).

2.4.6. Potential Ecological Risk Factor and Risk Index

Since PTEs have different abundances in the environment, sedimentation capacities, depending on the physicochemical conditions, and toxicities for aquatic biota, and the potential threat of PTEs for aquatic ecosystems also varies. To take these prerequisites into account, the adverse effect of individual PTEs was assessed by calculating the potential ecological risk factor (*Er*) and their total negative impact using the potential ecological risk index (*RI*) developed by Hakanson [42], according to:

$$Er = Tr_i \times CF_i, \tag{8}$$

$$RI = \sum_{i=1}^{n} Er_i, \tag{9}$$

where CF_i is the Contamination Factor and Tr_i is the toxic response factor for a given element (Table A3). Table A4 shows the classification of *Er* and *RI*.

2.5. Sediment Quality Assessment

The potential adverse impacts on benthos of pollutants from sediments have been assessed indirectly using SQGs. This approach is widely used to identify priority sediment pollutants and regions of concern [2,41,43,72].

In this work, the apparent effect threshold (AET) scheme and the US National Oceanic and Atmospheric Administration (NOAA) scheme were used as SQGs. The AET scheme includes the threshold effect level (*TEL*), which corresponds to the limit value at which the likelihood of adverse effects occurring is negligible, and probable effect level (*PEL*), which indicates the toxicity of the sediment and a high likelihood of adverse biological effects occurring. The NOAA scheme includes the effect range low (*ERL*), where adverse environmental effects are rare, and the effect range median (*ERM*), where adverse environmental effects are frequent. Concentrations > *ERL* and < *ERM* show irregular environmental responses [2,43]. The number of *TEL/ERL* and *TEL/ERM* exceedances indicates potential sediment toxicity. The SQGs are established for river and marine sediments depending on the sensitivity of aquatic organisms to pollution (Table A3).

The mean *ERM* quantity (*MERMQ*) scheme was used to estimate the overall adverse environmental effects of PTEs on aquatic sediments. The *MERMQ* method requires normalizing the concentration of each chemical element to its *ERM* value, summing the resulting coefficients, and dividing by the number of PTEs under study (10) and (11):

$$ERMQ = C_i / ERM_i, \tag{10}$$

$$MERMQ = \left(\sum_{i=1}^{n} ERMQ_i\right) / n, \tag{11}$$

where *ERM* is the effect median range for a given element (Table A3) and *n* is the number of elements analyzed. *MERMQ* ranges (Table A4) have been related to the probability of toxicity (76%, 49%, 21% and 9%, respectively) in amphipod assemblages.

However, since the *MERMQ* is not applicable for assessing the potential for adverse effects at relatively low concentrations of PTEs [40], the toxic risk index (*TRI*) proposed by Zhang et al. [2] was used. The integrated *TRI* for all analyzed PTEs was obtained by summing up the individual *TRIs*, calculated as follows:

$$TRI_i = \sqrt{\left([C_i/TEL]^2 + [C_i/PEL]^2\right)/2}, \tag{12}$$

$$TRI = \sum_{i=1}^{n} TRI_i \tag{13}$$

where *TEL* is the threshold effect level and *PEL* is the probable effect level (Table A3). The obtained values are classified into five toxic risk classes (Table A4).

2.6. Data Analysis and Visualization

Statistical analysis was carried out using STATISTICA 12 (StatSoft, Tulsa, OK, USA). The normality of the data was checked by the Kolmogorov–Smirnov and Lilliefors tests. Differences in sediment characteristics were assessed using one-way analysis of variance

(ANOVA) followed by Scheffe's post hoc test for normally distributed parameters that passed Levene's test, and Kruskal–Wallis ANOVA followed by multiple comparisons of mean ranks for all groups. Principal component analysis (PCA) was used to identify significant multi-element associations and the influence of the main physicochemical parameters of sediments on their geochemistry. Only components with eigenvalues > 1 according to the Kaiser criterion were considered.

Data visualization was conducted using Grapher 17 (Golden Software, Golden, CO, USA). Violin plots and ballon plots were generated with ggpubr R in the SRplot [77]. Maps illustrating the spatial distribution of the complex pollution and risk indices were created in Surfer 15 (Golden Software, Golden, CO, USA). Data interpolation was conducted using the inverse weighted distance (IWD) method with a power of 2. The basemaps were from Rosreestr and the satellite images were from Bing Maps.

3. Results and Discussion

3.1. Physicochemical Properties of Sediments

The main physicochemical properties of sediments of the study area that create conditions for the accumulation of PTEs were determined (Figure 2, Table A1). The distribution of pH and fine clay is close to normal, while EC and $CaCO_3$ have a pronounced right-skewed distribution, and the distribution of TOC and $\sum(Ca^{2+} + Mg^{2+})$ is multimodal. The pH values ranged between 6.5 and 9.5 (mean 7.4 ± 0.5). Most of the sediments (93%) were neutral or slightly alkaline, and only a few samples had pH values greater than 8, indicating moderate to strong alkalinity. Beach and shoreline sediments had higher pH on average (7.6 and 7.4, respectively) compared to delta sediments ($p < 0.005$). Riverbed and channel sediments within the delta were characterized by the greatest pH variability (Figure 2A). EC values ranged from 113 to 7680 μS/cm (mean 1523.5 ± 1504.9 μS/cm). A third of the samples were not saline, the majority were characterized by moderate salinity. On average, higher EC was noted in the bottom sediments of both the sea coast (2646.2 μS/cm) and the delta (1777.3 μS/cm). The content of $CaCO_3$ was from 0.4 to 27.2% (mean 6.1 ± 4.9%). Most samples were weakly to moderately carbonate, but coastal sediments had higher average $CaCO_3$ content (Figure 2B). TOC values were 0.1–0.5% (mean 1.5 ± 1.4%). Significantly lower TOC levels ($p < 0.0005$) were observed in beach sediments (0,4%), while higher TOC levels were recorded in stream sediments (2.8%). The sediments of the study area contain a fine clay fraction (<1 μm) between 1 and 36.3% (mean 13 ± 8.6%) (Figure 2C). The proportion of fine clay statistically significantly ($p < 0.005$) increases in shore sediments of the bay (18.3%) and in streamways (17.3%) in the delta. The distribution pattern of exchangeable cations is similar (Figure 2D). The content of $\sum(Ca^{2+} + Mg^{+2})$ varies from 2.2 to 46.3 cmol/kg (mean 21.0 ± 12.6 cmol/kg).

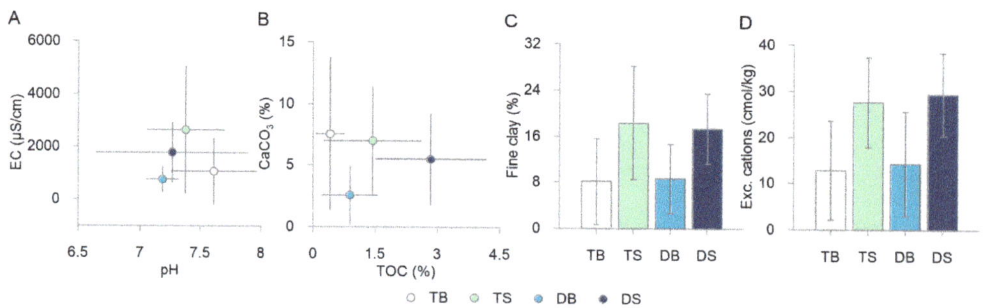

Figure 2. Variations in pH and electrical conductivity (EC) (**A**), total organic carbon (TOC) and $CaCO_3$ content (**B**), as well as content of fine clay (**C**) and exchangeable cations (**D**), expressed as mean values ± standard deviations, in beach and shore sediments of the Taganrog Bay (TB and TS, respectively) and in riverbank and stream sediments from the Don River delta (DB and DS, respectively).

Differences in the physicochemical properties of various types of sediments are due to hydrodynamic processes in the estuarine zone of the study region. Fine particles carried by river waters, which have the greatest sorption capacity, and organic matter with an affinity for the clay fraction are deposited in the channel facies of delta sediments [12,18,36]. Beach sand sediments on the coast of the bay are formed by the wave surf flow and gradually pass into coastal facies beyond the coastline, where the wave effect is less destructive [6,11], and the proportion of fine fractions is correspondingly higher.

3.2. Geochemistry of Sediments

The levels of PTE in the sediments of the study area are illustrated in Figure 3. The results showed that the distributions of Cr and Mn correspond to normal, but Cu, Zn and Cd are positively shifted, and Ni and Pb have a multimodal distribution in the sediments of the studied transition zone. The average content of PTEs in sediments of the study area decreases in the following order (in mg/kg): Mn (1272) > Cr (85.6) > Zn (85.5) > Ni (84.5) > Pb (76.3) > Cu (25.4) > Cd (2.1). Manganese, Cu, Zn and Cd reached the highest concentrations of 3516, 194, 481 and 8.5 mg/kg, respectively, in the beach sediments of the bay. The highest Pb contents (170 mg/kg) were recorded in the stream sediments of the delta, while the maximum contents of Cr (141 mg/kg) and Ni (151 mg/kg) were observed in the riverbank sediments. At the same time, the minimum contents of Mn (101 mg/kg) and Pb (18.2 mg/kg) were recorded sediments of the Don River delta, while the minimum concentrations of the remaining PTEs are characteristic of beach sediments of Taganrog Bay coast.

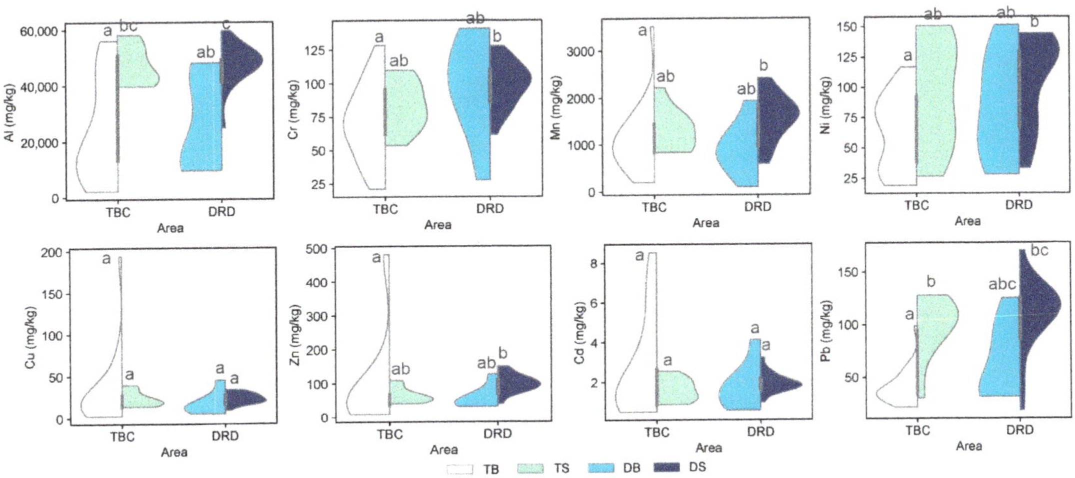

Figure 3. Violin plots showing PTE content in beach and shore sediments of the Taganrog Bay (TB and TS, respectively) and in riverbank and stream sediments from the Don River delta (DB and DS, respectively). Different letters indicate significant differences ($p < 0.05$) resulting from the multiple comparisons of p-values.

Hydrological regime and landscape conditions form the features of sedimentation in different environments [10,13,64,70]. Sediment characteristics such as pH, redox potential, texture and TOC influence the bioavailability of metals [2,26,36,38]. Thus, the sediments of riverbeds and channels of the Don River delta and shore sediments of the coast of the Taganrog Bay are characterized by a comparatively higher salinity, content of organic matter and exchangeable carbonates and heavier textures (Figure 2), which contribute to the absorption and accumulation of PTEs [44,78]. The levels of Cr, Mn, Ni and Zn in riverbed sediments are 1.1–1.7 times higher than in other types of sediments in the study area, and the average Pb content is up to 2.7 times higher (Figure 3).

The contents of Mn, Cd, Ni and Pb in the sediments of the estuarine zone of the Don River are 1.5–2 times higher than in the soils of the Lower Don floodplain and the coast of the Taganrog Bay, while the levels of Cr, Zn and Cu, on the contrary, are 1.1–1.7 times lower [45,46]. At the same time, a dilution effect is observed in the marine bottom sediments of the Taganrog Bay, which is manifested in a decrease of 2–5 times in the average contents of Mn, Ni, Pb and Cu compared to coastal sediments [55]. Thus, soils act as sources of elements of the first group, which are then intensively accumulated by coastal sediments.

The PTE levels in the Don River estuarine sediments are comparable to other sea–river transition zones in the world. In modern sediments of the Yellow River Delta, China, the Cu and Zn contents are 28.5 and 77.8 mg/kg, respectively [6], which is comparable to the Don River Delta. In sediments of the Ogun River Estuary, Nigeria, the average contents of Ni, Cd and Pb (22.2, 0.7 and 20.4 mg/kg, respectively) are 3–4 times lower than in the Don Estuary [5], which is due to lower urbanization and industrialization of the region. Lee et al. found lower levels of all PTEs in the 0–20 cm sediment layer of the Nakdong River Estuary, South Korea [15], than in the sediments of the study region. The coast of South Andaman Island, India, is characterized by higher average contents of Cr and Cu (390 and 48 mg/kg, respectively), comparable Ni and Zn (108 and 70 mg/kg, respectively), and lower levels of Mn and Pb (538 and 19 mg/kg, respectively) in sediments compared to the coast of Taganrog Bay [21]. The enrichment in Cr and Cu is probably due to the volcanic origin of the island's sediments. On the Mediterranean coast near Alexandria, Egypt, the average Zn content (160.4 mg/kg) is twice as high as in the Taganrog Bay, while the levels of other PTEs are 2–5 times lower. Mostafa et al. attribute Zn input to untreated municipal wastewater and oil refinery effluents, as well as transport emissions [17]. Thus, differences in the geochemistry of sea–river transition zone sediments are due to both natural and anthropogenic factors.

3.3. Potential Sources of PTEs

Environmental factors determine differences in the geochemistry of sediments along the fluvial-to-marine transition zone of the Don River. Factor 1 versus Factor 2 of PCA, using sediment characteristics and PTE concentrations, explains most of the data variability (61.2%). The distribution areas of different types of sediments of the study area overlap in the PCA space (Figure 4A). Seashore and floodplain sediments gravitate toward the centroid of the PCA space. At the same time, the stream sediments of the delta mainly occupy the space of positive values of Factor 1 and negative values of Factor 2; additionally, the coastal sediments of the bay are concentrated in the space of positive values of Factor 2. High and moderate positive correlations with Factor 1 are demonstrated by Cr (0.566), Mn (0.735), Ni (0.818), Pb (0.79). These PTEs are associated with TOC, fine clay, and exchangeable cations, which have moderate correlations with Factor 1 (0.658, 0.624 and 0.661, respectively). The remaining metals, namely Cu, Zn and Cd, are positively correlated with Factor 2 (0.637, 0.526 and 0.811, respectively). It is also worth noting that no significant correlations were found between pH, EC and carbonate content with any of the identified factors (Figure 4B).

The PCA results showed that Cr, Mn, Ni and Pb may originate from a common source and their behavior in sediments is controlled by the same conditions, whereas Cu, Zn and Cd may originate from a different source [2]. The accumulation of the Cr–Mn–Ni–Pb association in sediments is facilitated by an increase in the organic matter content and a heavier texture, most pronounced in the delta stream sediments as shown above (Figures 2 and 3). The flow of the Don River, which drains agriculturally developed plowed watersheds in its lower reaches [45,46], makes a significant contribution to the input of PTEs of this group into the sediments of the estuarine zone of the Don River [31,50,54]. Beach sediments of Taganrog Bay act as a geochemical barrier for Cu, Zn and Cd coming from surrounding terrestrial landscapes [45,59]. The origin of the Cu–Zn–Cd association in sediments is most likely associated local anthropogenic sources in the coastal area of the study region, along which urban and rural settlements are located [45]. Potential sources of PTEs include municipal and industrial wastewater, and transport and industrial emissions [56,57]. The significant contribution of anthropogenic sources for Cu, Zn and Cd in marine and deltaic sediments is also noted by other researchers [1,36,38].

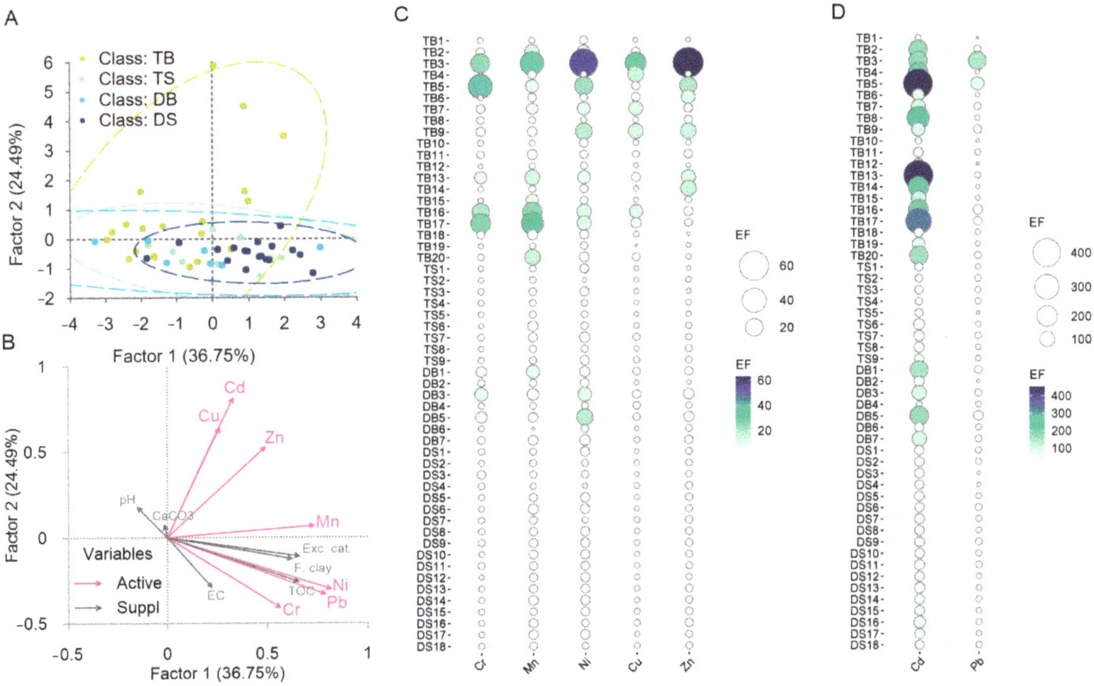

Figure 4. Biplots of the first two PCAs illustrating the sample points (**A**) and main physicochemical parameters and PTEs (**B**), as well as bubble plots of enrichment factors (*EF*) of PTE (**C**,**D**) in beach and shore sediments of the Taganrog Bay (TB and TS, respectively) and in riverbank and stream sediments from the Don River delta (DB and DS, respectively). Confidence ellipses ($\alpha = 0.05$) of the positions of group centroids are superimposed by dashed lines.

The contribution of anthropogenic sources to the accumulation of PTEs by sediments in the study area was assessed based on the *EF* values, which represent the excess of PTE contents in sediments over the UCC, normalized to the reference element. Extremely high enrichment was noted in beach sediments of the Taganrog Bay (Figure 4C,D). *EF* values up to 50 were found for Cu (29), Mn (37) and Cr (40); *EF* was > 50 for Ni (56), Zn (63), Pb (149) and Cd (453). Values 1 < *EF* < 3 were found for Cr (2.2), Zn (2.3) and Cu (2.6); 5 < *EF* < 10 for Mn (5.6) and Ni (6.1); 10 < *EF* < 25 for Pb (13) and 25 < *EF* < 50 for Cd (48) in shore sediments. The maximum *EF* values of riverbank and streamway sediments increase in the series Cu (3.7 and 1.8), Zn (6.1 and 3.5), Cr (10.2 and 3.8), Mn (10.3 and 5.6), Ni (18 and 5.1), Pb (36 and 17), Cd (167 and 56). In general, beach and floodplain sediments are more enriched in metals compared to bottom marine and river sediments of the study area.

All sediment samples from the studied region are extremely highly enriched in Cd (Figure 4D), with the proportion of extreme samples in above-water marine and river sediments reaching 75% and 57%, respectively. Several studies show that high Cd enrichment in marine sediments is typical for urbanized regions [5,17,21]. Also, bottom marine and river sediments are highly enriched in Pb (56% and 57%, respectively). Cd and Pb are definitely of anthropogenic origin [21], with average *EF* values in different types of sediments being 31–154 and 10–22, respectively. Enrichment of beach sediments of the Taganrog Bay from high to extreme was noted in 20–30% of samples by Cr, Mn, Ni, Cu and

Zn, in 50% of samples by Pb. High enrichment of Cr, Mn, Ni (14–29% of samples) and high to extreme enrichment of Pb (72% of samples) are observed in floodplain sediments.

3.4. PTE Contamination of Sediments and Associated Ecological Risks

The maximum *Igeo* values for Cr did not exceed 0 in the sediments of the study area (Figure 5A). The highest *Igeo*, up to 4.8, was achieved for Cd in the beach sediments of the Taganrog Bay. Also in this area, *Igeo* maxima were noted for Mn (1.7), Cu (1.5) and Zn (2.2). The highest *Igeo* for Ni (1.5) and Pb (2.1) were recorded in the shore sediments of the bay and the stream sediments of the delta, respectively. Among all the elements, Cd turned out to be the most dangerous: contamination of the sediments of the study area (% of samples) varied from weak (6%) and moderate (22% and 48%) to high (17%) and very high (7%) (Figure 5E). Pollution according to *Igeo* for Mn, Ni and Cu reached a moderate level in 1.9–35% of samples, for Zn and Pb moderate and considerable (from moderate to high according to Müller [74]) in 3.7% and 50% of samples, respectively.

Higher pollution estimates are obtained by *PI* values (Figure 5B) than from *Igeo*, since *PI* does not allow for the natural variability of elements to be taken into account [74]. The elements form the following series by decreasing maximum *PI* values: Cd (43) > Zn (6.8) > Pb (6.6) > Mn (4.9)> Ni = Cu (4.3) > Cr (1.5). Relatively higher *PI* values are noted for marine sediments compared to river sediments. High and very high sediment pollution by *PI* was obtained in 1.9–9.3% of samples for Mn, Cu and Zn, in 35% of samples for Ni, 50% of samples for Pb and 94% of samples for Cd (Figure 5F). The sediments of the estuarine zone of the Don River are more contaminated with Cr, Mn, Ni and Cu than the soils of the Lower Don floodplain and the coast of the Taganrog Bay [45]. Thus, marine and deltaic sediments of the estuarine zone of the Don River act as the final depot of elements in cascade coastal ecosystems [50].

The total pollution of sediments with PTE relative to the regional geochemical background was estimated by the *NPI* (Figure 6A). In the sediments of the study area, the *NPI* values varied within a range of 1.8–31, with an average of 7.8. The average *NPI* values in the beach and shore sediments of the Taganrog Bay were 9.7 and 6.0, and in the floodplain and channel sediments of the Don Delta they were 6.4 and 7.3, respectively. Pollution hotspots have been identified along the northern coast of Taganrog Bay. The majority of samples in the study area were assessed as highly contaminated (87%), while moderately and slightly contaminated accounted for 11% and 2%, respectively.

The elements in decreasing order of maximum *CF* values form the following series: Cd (36) > Zn (13) > Mn (10) > Pb (9.5) > Cu (6.5) > Ni (6.1) > Cr (3.4). The maximum *CFs* in beach sediments of the Taganrog Bay were recorded for Mn, Cu, Zn and Cd, in floodplain sediments for Cr and Ni and in delta beds for Pb (Figure 5C). River stream sediments were characterized by higher average *CF* values of all elements, except Cu and Cd, compared to other sediment types. Modern levels of Cr and Zn in sediments of the bottom estuarine zone exceed pre-industrial levels by two times, while Mn, Ni, Pb exceed them by three–four times and Cd by nine times. This fact indicates the anthropogenic nature of these changes [15,71]. The results obtained are consistent with the results of studies of the bottom sediments of the Azov Sea and Taganrog Bay, where an increase in the content of PTE has been noted over the past 30 years [50,51,57], which is a consequence of economic development, urbanization and industrialization of the Azov basin. According to the *CF*, the pollution varied from low to considerable for Cr, from low to very high for Mn, Ni, Cu and Zn and from moderate to very high for Cd and Pb (Figure 5G). It should be noted that the proportions of highly contaminated Cd and Pb samples were 67% and 33%, respectively.

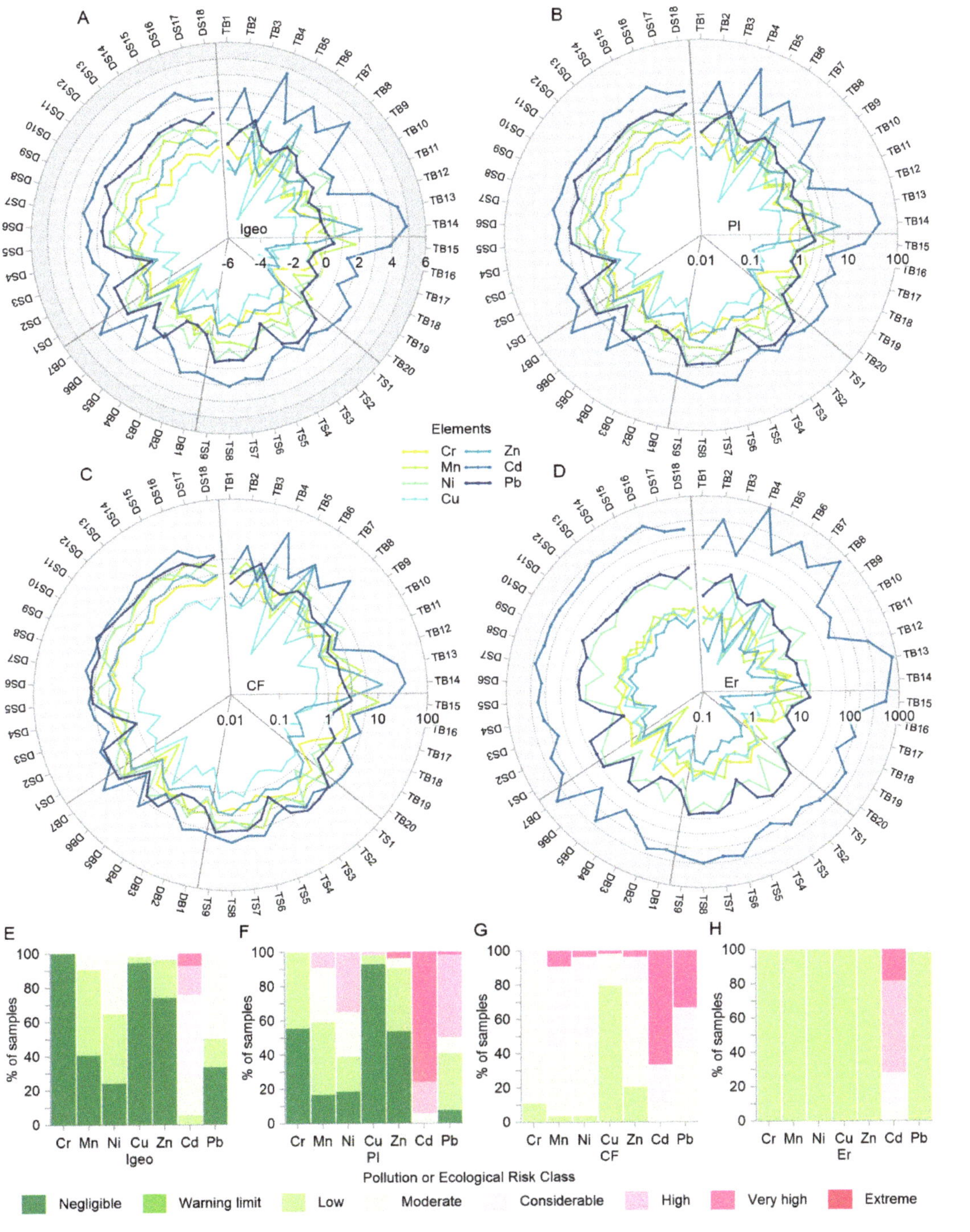

Figure 5. Values of geoaccumulation indices (*Igeo*) (**A**), single pollution indices (*PI*) (**B**), contamination factors (*CF*) (**C**) and potential ecological risk factors (*Er*) (**D**) in sediments of the study area and the corresponding distribution of samples by pollution (**E–G**) and ecological risk classes (**H**).

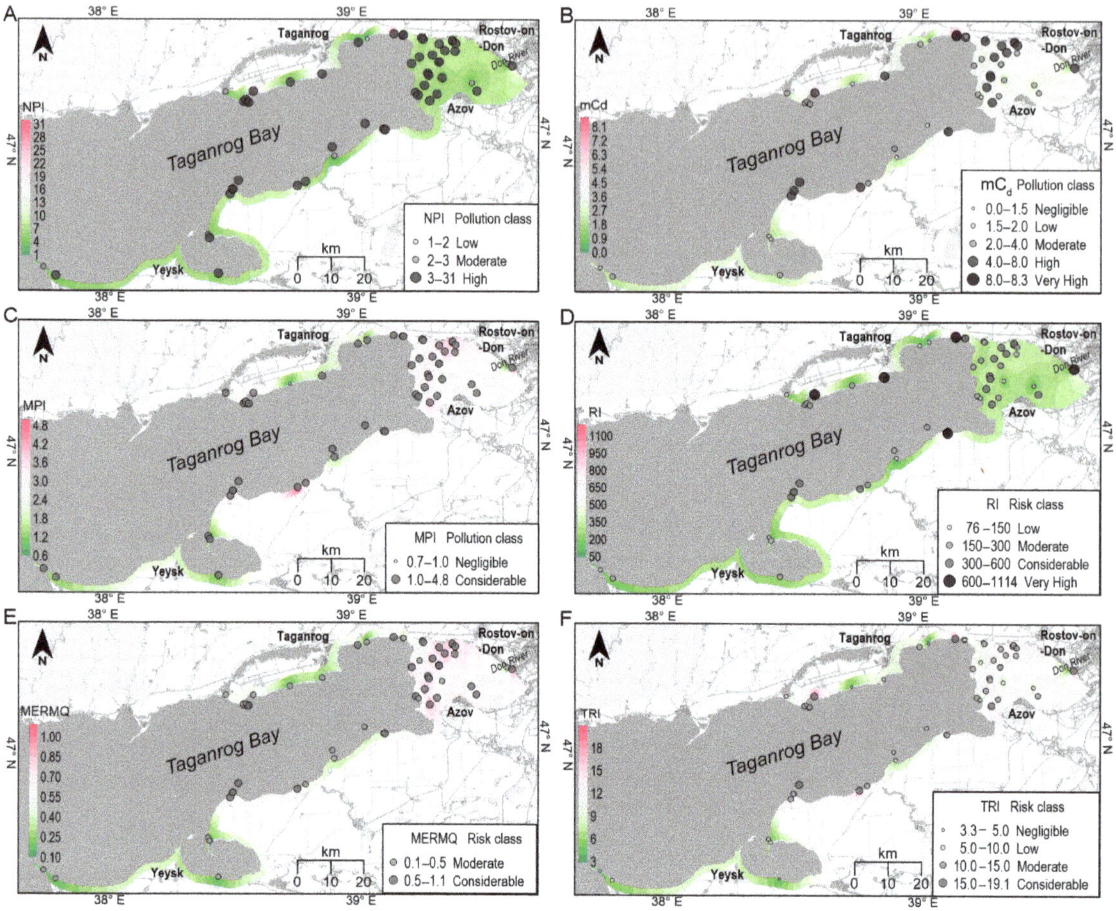

Figure 6. Spatial distribution of the Nemerow pollution index (*NPI*) (**A**), modified pollution degree (mC_d) (**B**), metal pollution index (*MPI*) (**C**), potential ecological risk index (*RI*) (**D**), mean effect range median quantity (*MERMQ*) (**E**) and integrated toxic risk index (*TRI*) (**F**) in sediments of the study area, as well as classification of samples by total PTE pollution (**A–C**) and ecological risk classes (**D–F**).

Relative to the pre-industrial level, the overall pollution was assessed using mC_d and *MPI*. The mC_d values varied within 1–8 (mean 3.6), and *MPI* ranged within 0.8–4.7 (mean 2.7). According to both mC_d and *MPI*, only 3.7% of sediment samples from the study area are classified as unpolluted (Figure 6B,C). The mC_d analysis allowed us to obtain a more detailed assessment of the overall pollution level. Thus, the shares of slightly and moderately polluted sediments were 13.0% and 44.4%, respectively, while those of highly and very highly polluted sediments were 37.0% and 1.9%, respectively. An increased level of overall pollution with metals was noted in sediments of the Don River delta (Figure 6B).

Analysis of potential ecological risks of individual elements showed that *Er* values for Cr, Mn, Ni, Cu and Zn did not exceed 40; *Er* for Pb reached 47 in only one sample (Figure 5D). Accordingly, most of the studied PTEs pose a low individual potential ecological risk and make a minimal contribution to the overall risk (Figure 5H). *Er* values for Cd varied within 59–1068, with an average value of 21.2, reaching a maximum in the beach sediments of the bay. In general, Cd levels in two thirds of the samples of different types in the study area pose a high and very high potential ecological risk. Similar assessments of the ecological risk of all elements except Cd were obtained for the soils of the Lower Don and the coast of the Taganrog Bay [46].

RI values varied significantly from 76 to 1114 in the sediments of the study area. An extremely high overall risk was noted in the coastal sediments of the Taganrog Bay near residential areas (Figure 6D). The total pollution of sediments of the study area by PTE presents a low ecological risk in 20.4% of samples, moderate and considerable risk in 37.0% and 33.3% of samples, respectively, and high risk in 9.3% of samples.

According to individual pollution indices of *Igeo*, *PI* and *CF*, the priority pollutants of sediments in the study area were Cd, Pb and Zn. Moreover, high levels of these elements also pose the greatest ecological risk (Figure 5). The leading role of Cd in sediment contamination and contribution to ecological risk has also been noted near the mouths of the Guarda River and Ita Channel, Brazil [38], in the wetlands of the Yellow River delta, China [2], along the coast of East Java, Indonesia [72], and on the Egyptian Mediterranean Sea coast [22].

3.5. Ecotoxicology of Sediments

The sediment samples from Taganrog Bay do not exceed the *ERMs* for Cr, Cu, Cd and Pb (Table 1). Most of the Cr, Cu and Pb contents are also below the *ERL* values, while Cd is mainly in the range from *ERL* to *ERM*. Although most Zn values are below the *ERL*, in individual samples the contents exceed the *ERM*. Only Cr does not exceed the *PEL* in all sediments on the coast of the bay. Most of the sediment samples are below the *TEL* for Cu and Zn, and in the range from *TEL* to *PEL* for Cd and Pb. Ni deserves priority attention, the levels of which in most samples exceeded the *ERM* and *PEL*. Thus, Cr and Cu pose a relatively low ecotoxic risk in the sediments of the Taganrog Bay, while Zn, Cd and Pb pose a moderate risk, and Ni poses a high risk. The importance of Ni as a pollutant of particular interest has also been noted in coastal marine sediments of the Mediterranean Sea [22]. The sediment samples from the Don River delta do not exceed the *ERM/PEL* for Cu and Zn; in most samples, their levels are below the *ERL/TEL* (Table 1). Thus, these elements pose a minimal risk of adverse effects. The contents of Cr and Cd in most samples are below the *ERM*, but above the *TEL*, indicating moderate risks. The main high-risk sources are Ni and Pb, the levels of which in most delta sediment samples exceed the *ERM/PEL*. These results are consistent with previous studies showing that the bottom sediments of the Lower Don were moderately contaminated with Ni [31].

Table 1. Relative percentage of samples amongst ranges of Sediment Quality Guidelines.

Area	Elements	Effects Range			Effects Level		
		<*ERL*	*ERL–ERM*	≥*ERM*	<*TEL*	*TEL–PEL*	≥*PEL*
Taganrog Bay	Cr	62.1	37.9	0	17.2	82.8	0
	Ni	3.4	27.6	69	0	31	69
	Cu	79.3	20.7	0	51.7	44.8	3.4
	Zn	89.7	3.4	6.9	86.2	6.9	6.9
	Cd	37.9	62.1	0	10.3	72.4	17.2
	Pb	55.2	44.8	0	17.2	72.4	10.3
Don River Delta	Cr	20	80	0	4	24	72
	Ni	4	12	84	0	12	88
	Cu	100	0	0	96	4	0
	Zn	80	20	0	84	16	0
	Cd	100	0	0	4	92	4
	Pb	16	36	48	16	16	68

In descending order of maximum *ERMQ* values, the elements formed the following series: Ni (3) > Pb (1.5) > Zn (1.2) > Cr (1.0) > Cd (0.9) > Cu (0.7). *ERMQ* analysis showed that, on average, the risks of most elements, except Cu and Cd, were higher in the Don River Delta sediments compared to the Taganrog Bay coast (Figure 7A). According to

ERMQ, Cr, Cu, Zn and Cd demonstrated low to significant risks, while the proportions of samples with moderate and significant risks were 96%, 17%, 82% and 100%, respectively. High risk of Ni and Pb was found in 50% and 1.9% of sediment samples in the study area (Figure 7B). *MERMQ* values ranged from 0.14 to 1.02, with an average of 0.5. Moderate weighted average risk was characteristic of 55.6% of marine and river sediment samples from the estuarine region of the Don River, mainly in its delta (Figure 6E).

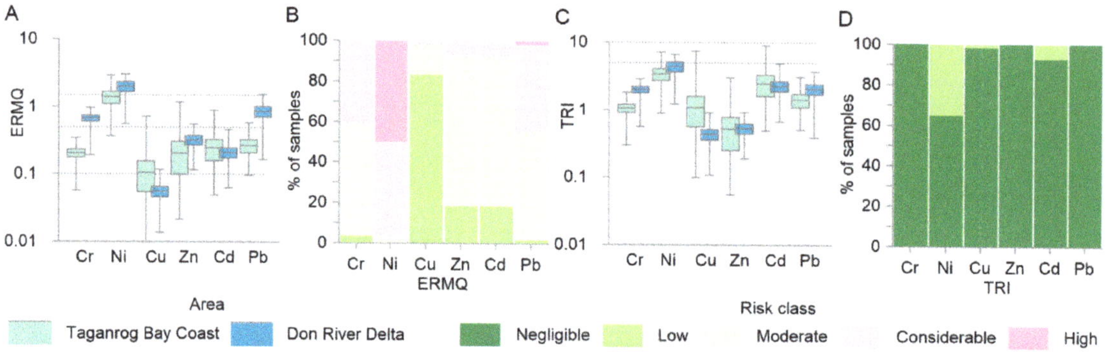

Figure 7. Variability of effect range median quantity (*ERMQ*) (**A**) and toxic risk indices (*TRI*) (**C**) in sediments of the study area presented as a box (mean ± 95% CI) and whisker (minimum–maximum) plot, as well as the percentage of distribution of individual ecological risks (**B**,**D**).

In all sediment samples, the *TRI* values were less than five for Cr, Zn and Pb (Figure 7C). The maximum *TRI* values for Ni, Cu and Cd were 7.2, 7.4 and 9, respectively. According to the *TRI*, a low risk of potential impact of these elements was found in 35%, 2% and 7% of the samples, respectively (Figure 7D). The values of the total toxic risk varied within the range of 3.3–19.1, with an average of 10.8. The total risk was classified from negligible and low to moderate and considerable (9.3%, 31.5%, 46.3% and 13%, respectively). In two thirds of the samples, it was above the acceptable low level. In general, the area of increased total toxic risks of sediment pollution by PTEs was observed in the Don River Delta (Figure 6F).

4. Conclusions

In the present study, we investigated the geochemistry, controlling factors, possible sources, pollution hazards and potential ecological and ecotoxicological risks of exposure to seven toxic elements (i.e., Cr, Mn, Ni, Cu, Zn, Cd and Pb) in sediments of the Don River estuarine region.

The average PTE content showed a decreasing order of Mn > Cr > Zn > Ni > Pb > Cu > Cd and was consistent with PTE levels in sediments of highly urbanized and industrialized river-sea transition zones. The results showed that the hydrological regime, landscape–geochemical conditions and anthropogenic impact determine the intensity of PTE accumulation by sediments of the study area. In particular, stream sediments of the Don River delta are characterized by higher levels of Cr, Mn, Ni, Zn and Pb on average than other types of sediments.

The origin of all the studied PTE is due to human activity, as shown by the PCA and *EF*. At the same time, it should be emphasized that Cr, Mn, Ni and Pb enter mainly with river runoff, and their behavior is controlled by the texture of sediments and the content of organic matter, while Cu, Zn and Cd are associated with local anthropogenic sources.

Based on the results of the analysis of *Igeo*, *PI* and *CF*, it was revealed that the priority pollutants of the sediments of the Don Delta and the coast of the Taganrog Bay are Cd and Pb. These elements make the greatest contribution to pollution and increased ecological risks. An increased level of total pollution of sediments in the Don River delta and along the coast of the Taganrog Bay, identified using the *NPI*, *MPI* and mC_d, was found locally near residential areas. A holistic approach to assessing the state of sediments in the study area included an analysis of ecotoxicological risks of potential impact of inorganic pollutants. It has been established that although the average level of PTE pollution of sediments in the Don River delta is lower than that of the surrounding coast, it poses a more significant risk of negative effects on freshwater organisms.

The study has a number of limitations, primarily due to the need for a detailed examination of the mechanisms of geochemical transformation and migration of elements in accessible forms between the biota and abiotic components of aquatic landscapes. Nevertheless, the preliminary results of the comprehensive assessment of sediment pollution, potential ecological and ecotoxicological risks of PTEs may be useful for improving environmental monitoring of the transit zone of the Don River–Sea of Azov system and achieving sustainable development goals.

Author Contributions: Conceptualization, T.M.; Data curation, T.B. and S.M.; Formal analysis, E.K.; Funding acquisition, T.M.; Investigation, I.Z., E.L., T.D. and M.B.; Methodology, E.K.; Project administration, S.M.; Resources, D.N. and M.B.; Supervision, T.M.; Validation, D.N., T.B. and R.K.Y.; Visualization, E.K.; Writing—original draft, E.K.; Writing—review and editing, T.M. and R.K.Y. All authors have read and agreed to the published version of the manuscript.

Funding: The study was supported by the grant from the Russian Science Foundation (project No. 20-14-00317) at the Southern Federal University.

Data Availability Statement: Data are openly available within this article in Appendix A.

Conflicts of Interest: The authors declare no conflicts of interest. The funders had no role in the design of the study; in the collection, analyses, or interpretation of data; in the writing of the manuscript; or in the decision to publish the results.

Appendix A

Table A1. Location and physicochemical properties of sediment samples from the study area.

Area	Site		Coordinates		pH	EC	TOC	$CaCO_3$	Fine Clay	$\Sigma(Ca^{2+} + Mg^{2+})$
						µS/cm	%			cmol/kg
Taganrog Bay coast	Beach	TB1	N47°08′43.26″	E38°29′02.69″	8.0	895	0.5	1.3	22.1	27.8
		TB2	N46°52′48.04″	E38°30′00.47″	7.5	184	1.1	5.7	11.0	17.3
		TB3	N47°06′54.36″	E38°34′08.69″	7.8	417	0.1	5.5	1.5	3.1
		TB4	N47°08′37.10″	E38°35′37.10″	8.1	142	0.5	1.4	13.3	21.8
		TB5	N47°09′53.32″	E38°44′34.04″	7.2	336	0.3	6.1	4.7	6.1
		TB6	N46°53′02.08″	E38°45′34.99″	7.8	113	0.2	12.7	16.4	22.8
		TB7	N46°53′41.78″	E38°47′29.00″	7.4	173	0.3	6.1	4.6	5.5
		TB8	N47°11′02.98″	E38°52′17.58″	7.5	237	0.2	1.8	2.9	4.2
		TB9	N46°57′45.83″	E38°54′31.32″	7.7	167	0.1	2.3	2.9	7.9
		TB10	N47°16′03.11″	E39°01′04.04″	7.8	318	0.7	4.1	20.5	26.0
		TB11	N47°16′33.64″	E39°03′20.70″	7.3	668	0.4	2.9	6.1	7.1
		TB12	N47°01′43.39″	E39°06′33.84″	8.4	116	0.1	5.0	24.0	41.5
		TB13	N47°01′36.91″	E39°06′54.43″	7.6	354	0.6	2.2	6.7	10.9
		TB14	N47°17′11.18″	E39°09′50.18″	7.0	856	0.3	11.0	11.0	19.6
		TB15	N47°16′56.50″	E39°12′03.67″	7.1	805	1.4	9.1	7.0	15.5
		TB16	N46°41′01.21″	E37°44′11.15″	7.6	3880	0.1	11.2	1.5	4.5
		TB17	N46°39′32.98″	E37°47′05.68″	7.9	3248	0.4	12.3	1.1	4.1
		TB18	N46°45′32.40″	E38°23′54.78″	7.3	3080	0.2	27.2	1.0	2.7
		TB19	N46°45′02.84″	E38°24′16.49″	7.4	2884	0.4	14.2	3.4	5.5
		TB20	N46°39′08.10″	E38°26′03.44″	7.8	2312	0.2	9.3	1.3	3.1

Table A1. *Cont.*

Area		Site	Coordinates		pH	EC	TOC	$CaCO_3$	Fine Clay	$\Sigma(Ca^{2+} + Mg^{2+})$
						µS/cm	%			cmol/kg
	Shore	TS1	N47°07′02.50″	E38°33′15.05″	7.5	1258	0.4	3.8	15.1	22.1
		TS2	N47°07′13.66″	E38°33′45.47″	7.2	2750	1.3	17.1	14.4	17.8
		TS3	N47°06′51.84″	E38°34′24.85″	7.9	1740	1.5	7.1	19.3	26.6
		TS4	N46°59′08.77″	E38°54′06.95″	7.2	1193	0.9	5.2	9.2	21.8
		TS5	N47°02′46.79″	E39°02′02.87″	7.3	888	1.1	5.2	8.0	19.7
		TS6	N47°16′57.43″	E39°12′06.77″	7.1	1166	1.5	7.2	8.9	22.6
		TS7	N46°52′01.81″	E38°29′22.45″	7.6	5700	1.6	9.5	36.3	42.4
		TS8	N46°52′48.83″	E38°30′03.46″	7.7	1441	0.3	6.0	25.2	31.5
		TS9	N46°54′14.80″	E38°31′23.52″	6.9	7680	4.3	2.0	28.3	44.0
Don River Delta	Riverbank	DB1	N47°16′22.77″	E39°16′29.42″	7.3	650	0.6	7.0	4.7	9.5
		DB2	N47°12′17.93″	E39°20′17.59″	7.4	1381	0.7	2.5	13.2	19.7
		DB3	N47°09′43.85″	E39°20′48.05″	7.1	321	0.4	0.5	2.0	3.3
		DB4	N47°08′40.88″	E39°27′53.71″	7.0	876	1.7	2.2	14.2	28.4
		DB5	N47°07′18.26″	E39°28′55.45″	7.2	487	0.3	0.4	3.5	2.2
		DB6	N47°11′30.59″	E39°36′42.52″	7.2	206	0.5	1.0	5.6	8.1
		DB7	N47°11′10.72″	E39°37′46.02″	7.1	1356	2.0	4.2	17.1	28.7
	Stream	DS1	N47°12′11.34″	E39°13′46.96″	6.5	3100	3.5	4.1	13.6	20.6
		DS2	N47°07′27.73″	E39°14′31.20″	7.5	1527	1.7	8.1	20.2	31.1
		DS3	N47°07′04.91″	E39°15′08.06″	7.0	1298	0.9	1.9	13.0	27.8
		DS4	N47°13′12.00″	E39°15′44.35″	7.2	145	0.4	0.4	5.8	4.4
		DS5	N47°08′33.36″	E39°16′40.40″	7.3	1467	1.4	4.0	20.2	29.4
		DS6	N47°16′08.15″	E39°16′45.01″	7.5	1882	2.8	6.3	14.6	27.8
		DS7	N47°10′33.78″	E39°17′32.39″	7.4	2570	3.8	5.3	31.5	46.3
		DS8	N47°10′12.14″	E39°17′44.66″	7.0	2170	2.4	3.1	23.0	32.1
		DS9	N47°06′02.77″	E39°17′45.42″	6.6	2590	2.4	8.0	13.4	27.2
		DS10	N47°13′27.59″	E39°18′44.82″	7.5	1110	2.8	7.9	17.3	29.1
		DS11	N47°07′04.87″	E39°19′27.80″	7.2	234	2.4	2.7	14.1	26.1
		DS12	N47°12′14.80″	E39°20′17.16″	7.0	1456	2.2	3.9	15.3	27.3
		DS13	N47°15′14.94″	E39°20′30.70″	7.3	1357	4.4	8.3	15.1	26.0
		DS14	N47°14′10.86″	E39°21′41.51″	7.0	1431	4.5	5.5	13.8	26.6
		DS15	N47°15′39.17″	E39°23′36.78″	7.1	1130	4.2	6.8	29.5	43.6
		DS16	N47°15′45.97″	E39°23′38.58″	7.0	2010	4.3	16.9	18.2	38.4
		DS17	N47°14′00.89″	E39°24′08.57″	7.2	1324	2.0	2.6	14.6	28.1
		DS18	N47°15′14.72″	E39°24′21.85″	9.5	5190	5.0	3.0	18.9	36.0

Table A2. Concentration of PTEs (mg/kg) in sediment samples from the study area.

Site	Al	Cr	Mn	Ni	Cu	Zn	Cd	Pb
TB1	56,074	109.6	871.2	102.5	19.9	43.5	0.7	33.1
TB2	14,812	66.2	1761.6	81.9	18.3	25.9	2.7	59.1
TB3	3174	85.9	1123.9	102.3	32.2	164.2	0.6	98.5
TB4	35,972	40.9	1327.4	38.0	194.0	47.2	7.6	39.9
TB5	2116	95.3	194.6	26.3	2.6	31.5	1.1	33.8
TB6	52,371	74.3	1826.0	88.7	73.7	481.4	3.5	59.2
TB7	16,399	77.5	1127.0	92.2	68.9	51.7	1.6	55.3

Table A2. *Cont.*

Site	Al	Cr	Mn	Ni	Cu	Zn	Cd	Pb
TB8	20,116	62.4	417.6	19.1	9.4	53.3	5.5	34.9
TB9	7406	37.1	353.6	77.3	35.0	88.7	0.6	36.8
TB10	51,313	101.0	826.3	116.7	27.0	57.7	0.8	39.4
TB11	11,638	37.5	573.3	23.9	9.2	34.4	0.5	25.3
TB12	51,842	62.7	827.1	73.2	20.4	40.2	2.6	33.7
TB13	13,225	128.3	1731.9	74.8	19.7	127.4	6.5	32.9
TB14	38,088	62.3	801.1	74.1	16.5	461.9	8.5	38.7
TB15	40,204	58.1	3515.9	72.3	16.9	68.1	4.3	53.2
TB16	3703	81.2	881.1	27.8	14.0	13.1	0.8	24.9
TB17	3703	108.3	1171.4	26.4	7.3	8.8	1.4	21.2
TB18	21,160	74.2	1107.6	53.0	10.3	13.0	0.7	36.9
TB19	23,276	31.9	401.2	38.1	4.8	19.5	1.7	32.2
TB20	9522	21.3	1347.7	28.3	10.0	10.5	1.6	29.4
TS1	51,842	73.2	967.7	129.1	19.2	39.8	1.4	97.5
TS2	44,436	67.9	869.1	26.4	18.2	57.4	1.0	111.9
TS3	50,784	105.3	1447.3	130.9	22.4	49.6	1.4	127.8
TS4	42,320	84.9	833.3	52.6	15.7	53.2	0.8	94.4
TS5	39,675	69.0	857.5	52.8	13.8	43.3	1.1	29.7
TS6	41,791	53.4	2220.0	70.2	18.5	64.2	2.2	76.2
TS7	41,262	100.3	1626.4	145.3	36.9	64.6	2.0	113.6
TS8	57,132	87.0	1425.3	92.7	27.6	109.0	2.6	109.3
TS9	58,190	109.8	1469.8	150.8	39.7	96.3	2.0	121.6
DB1	10,051	69.1	986.5	33.9	10.0	31.2	1.6	41.3
DB2	45,494	136.0	1303.9	115.9	16.3	35.9	2.0	31.1
DB3	9522	110.1	667.1	63.2	12.3	47.4	0.9	32.0
DB4	48,139	94.5	945.3	68.6	45.7	75.6	0.6	92.7
DB5	11,638	97.2	529.6	117.6	10.9	53.7	2.1	86.2
DB6	16,399	27.6	101.0	28.0	5.4	31.1	0.6	33.4
DB7	41,262	140.8	1936.1	151.3	25.7	127.0	4.1	125.4
DS1	46,023	104.3	957.9	126.8	17.2	86.6	1.6	118.2
DS2	47,610	85.7	1124.0	69.0	21.6	84.8	1.9	117.3
DS3	24,863	105.5	587.6	43.7	9.5	39.4	1.0	18.2
DS4	46,552	118.9	738.0	33.0	25.1	100.6	1.6	49.8
DS5	47,610	91.4	1508.2	128.4	21.6	108.4	1.9	117.0
DS6	42,320	77.0	2245.2	124.8	17.1	83.4	1.6	127.9
DS7	55,545	117.2	1734.1	60.1	32.4	126.6	2.3	127.6
DS8	49,197	106.1	1591.1	132.5	24.2	103.8	1.9	108.4
DS9	49,197	127.8	1869.9	134.0	28.6	110.5	2.0	170.4
DS10	49,197	79.9	1830.2	126.9	19.6	87.7	1.9	122.9
DS11	37,559	102.2	1217.0	88.6	18.6	70.0	1.7	105.2
DS12	48,139	61.4	1597.3	128.9	21.3	77.1	1.8	94.1
DS13	51,842	98.4	1739.3	134.5	27.1	120.9	2.5	107.1
DS14	51,842	100.1	2423.3	144.3	27.4	109.9	2.1	125.1
DS15	59,777	119.1	1533.6	142.2	34.4	145.7	2.6	125.3
DS16	52,371	98.7	2178.7	87.3	32.7	150.5	3.2	128.9
DS17	43,907	80.7	1907.0	126.7	18.5	85.3	1.9	86.6
DS18	52,900	107.9	1520.0	66.7	29.0	107.0	1.8	126.7

Table A3. Parameters used to assess sediment pollution and ecological risk of PTEs.

Parameter		Unit	Cr	Mn	Ni	Cu	Zn	Cd	Pb	Reference
Geochemical background	UCC	mg kg^{-1}	92	774	47	28	67	0.09	17	[39]
	C_b		95	720	35	45	71	0.2	26	[78]
	C_0		41	347	25	30	37.4	0.24	18	[50,57]
Toxic response factor	*Tr*	unitless	2	1	5	5	1	30	5	[38,42]
Sediment quality guidelines *	TEL_f	mg kg^{-1}	37.3	–	18	35.7	123	0.596	35	[79,80]
	ERL_f		80	–	30	70	120	5	35	[79]
	PEL_f		90	–	36	197	315	3.53	91.3	[79,80]
	ERM_f		145	–	50	390	270	9	110	[79]
	TEL_s		52.3	–	15.9	18.7	124	0.68	30.24	[35,80]
	ERL_s		81.8	–	20.9	34	150	1.2	46.7	[80,81]
	PEL_s		160	–	42.8	108	271	4.21	112	[35,80]
	ERM_s		370	–	51.6	270	410	9.6	218	[80,81]

Note(s): * The subscripts *f* and *s* denote the effect levels for freshwater and marine sediments, respectively.

Table A4. Criteria for classifying metal sediment enrichment, pollution and ecological risks.

Index		Class								Reference
		Negligible	Warning Limit	Low	Moderate	Considerable	High	Very High	Extreme	
Enrichment	*EF*	≤1	–	1–3	3–5	5–10	10–25	25–50	>50	[40]
Pollution	*Igeo* *	≤0 (0)	–	0–1 (1)	1–2 (2)	2–3 (3)	3–4 (4)	4–5 (5)	>5 (6)	[74]
	PI	≤1	–	1–2	2–3	–	3–5	>5	–	[70]
	NPI	≤0.7	0.7–1	1–2	2–3	–	>3	–	–	[75]
	CF	–	–	<1	1–3	3–6	–	≥6	–	[42]
	mC_d	<1.5	–	1.5–2	2–4	–	4–8	8–16	16–32	[71]
	MPI	<1	1	–	–	>1	–	–	–	[76]
Ecological Risk	*Er*	–	–	<40	40–80	80–160	160–320	≥320	–	[42]
	RI	–	–	<150	150–300	300–600	–	≥600	–	[42]
	ERMQ	–	–	<0.1	0.1–0.5	0.5–1.5	≥1.5	–	–	[43]
	MERMQ	–	–	<0.1	0.1–0.5	0.5–1.5	≥1.5	–	–	[43]
	TRI	≤5	–	5–10	10–15	15–20	–	>20	–	[2]

Note(s): * *Igeo* classes according to Müller [74] are indicated in brackets.

References

1. Silva, M.C.; do Nascimento Monte, C.; de Souza, J.R.; Selfe, A.C.C.; Ishihara, J.H. Mapping of Metals Contamination in Coastal Sediments around the World in the Last Decades: A Bibliometric Analysis and Systematic Review. *Mar. Pollut. Bull.* **2024**, *205*, 116572. [CrossRef] [PubMed]
2. Zhang, G.; Bai, J.; Zhao, Q.; Lu, Q.; Jia, J.; Wen, X. Heavy Metals in Wetland Soils along a Wetland-Forming Chronosequence in the Yellow River Delta of China: Levels, Sources and Toxic Risks. *Ecol. Indic.* **2016**, *69*, 331–339. [CrossRef]
3. Bianchi, T.S.; Allison, M.A. Large-River Delta-Front Estuaries as Natural "Recorders" of Global Environmental Change. *Proc. Natl. Acad. Sci. USA* **2009**, *106*, 8085–8092. [CrossRef] [PubMed]
4. Sedakov, R.; Osadchiev, A.; Barnier, B.; Molines, J.M.; Colombo, P. Large Chocked Lagoon as a Barrier for River–Sea Flux of Dissolved Pollutants: Case Study of the Azov Sea and the Black Sea. *Mar. Pollut. Bull.* **2023**, *187*, 114496. [CrossRef]
5. Phillips, O.A.; Fajemila, O.T. Contamination Levels of Potentially Toxic Elements within the Ogun River Estuary Sediments, Southwest Nigeria: Ecological and Human Health Risk Assessments. *J. Trace Elem. Miner.* **2024**, *8*, 100120. [CrossRef]
6. Meng, L.; Wang, L.; Wang, Q.; Zhao, J.; Zhang, G.; Zhan, C.; Liu, X.; Cui, B.; Zeng, L. Geochemical Characteristics of the Modern Yellow River Delta Sediments and Their Response to Evolution of the Sedimentary Environment. *Front. Mar. Sci.* **2024**, *11*, 1370336. [CrossRef]
7. Liu, D.; Lin, Y.; Zhang, T.; Huang, E.; Zhu, Z.; Jia, L. Impact of Anthropogenic Activities on Sedimentary Records in the Lingdingyang Estuary of the Pearl River Delta, China. *J. Mar. Sci. Eng.* **2024**, *12*, 1139. [CrossRef]
8. Yuan, X.; Yang, Q.; Meadows, M.E.; Luo, X.; Wang, Z. Grain Size and Organic Geochemistry of Recent Sediments in Lingding Bay, Pearl River Delta, China: Implications for Sediment Dispersal and Depositional Processes Perturbed by Human Activities. *Anthr. Coasts* **2021**, *4*, 147–167. [CrossRef]

9. Zhang, J.; Liu, C.L. Riverine Composition and Estuarine Geochemistry of Particulate Metals in China—Weathering Features, Anthropogenic Impact and Chemical Fluxes. *Estuar. Coast. Shelf Sci.* **2002**, *54*, 1051–1070. [CrossRef]
10. Fedorova, I.; Chetverova, A.; Bolshiyanov, D.; Makarov, A.; Boike, J.; Heim, B.; Morgenstern, A.; Overduin, P.P.; Wegner, C.; Kashina, V.; et al. Lena Delta Hydrology and Geochemistry: Long-Term Hydrological Data and Recent Field Observations. *Biogeosciences* **2015**, *12*, 345–363. [CrossRef]
11. Mandour, A.S.; Ghezzi, L.; Lezzerini, M.; El-Gamal, A.A.; Petrini, R.; Elshazly, A. Geochemical Characterization of Recent Nile Delta Inner Shelf Sediments: Tracing Natural and Human-Induced Alterations into a Deltaic System. *Egypt. J. Aquat. Res.* **2020**, *46*, 355–361. [CrossRef]
12. Nascimento, S.C.; Parbhakar-Fox, A.; Cracknell, M.J.; Cooke, D.R.; Miller, C.B.; Heng, W.X. Geochemical, Mineralogical, and Geophysical Methods to Establish the Geoenvironmental Characteristics of the King River Delta, Queenstown, Western Tasmania. *Appl. Geochem.* **2023**, *159*, 105820. [CrossRef]
13. Passalacqua, P.; Lanzoni, S.; Paola, C.; Rinaldo, A. Geomorphic Signatures of Deltaic Processes and Vegetation: The Ganges-Brahmaputra-Jamuna Case Study. *J. Geophys. Res. Earth Surf.* **2013**, *118*, 1838–1849. [CrossRef]
14. Duţu, L.; Secrieru, D.; Duţu, F.; Lupaşcu, N. Geochemical Dataset of the Danube Delta Sediments. *Data Brief* **2021**, *39*, 107529. [CrossRef]
15. Lee, G.; Jung, N.; Dellapenna, T.; Ra, K.; Chang, J.; Kong, G.S.; Nahm, W.-H.; Park, B.S.; Jeong, H. Pace of Heavy Metal Pollution in the Anthropogenically Altered and Industrialized Nakdong River Estuary, South Korea: Implications for the Anthropocene. *Mar. Pollut. Bull.* **2024**, *205*, 116678. [CrossRef]
16. Berdnikov, S.V.; Sorokina, V.V.; Kleshchenkov, A.V.; Tyutyunov, Y.V.; Kulygin, V.V.; Kovaleva, G.V.; Bulysheva, N.I. Marine Indicators of Climate Change in the Azov Sea Ecosystem. *J. Sea Res.* **2023**, *193*, 102373. [CrossRef]
17. Mostafa, M.T.; El-Nady, H.; Gomaa, R.M.; Salman, S.A.; Khalifa, I.H. Evaluation of Urbanization Influences on Beach Sediment Contamination with Heavy Metals Along the Littoral Zone of Alexandria City, Egypt. *Water Air Soil Pollut.* **2024**, *235*, 759. [CrossRef]
18. Jolivet, M.; Dauteuil, O.; Dia, A.; Davranche, M.; Pierson-Wickmann, A.-C.; Barrier, L.; Murray-Hudson, M.; Mazrui, N.; Marsac, R.; Cheng, F.; et al. Highly Contrasted Geochemical Pattern in Sediments of the Okavango Delta, Botswana Driven by Dust Supply, Hydrological Heritage and Biogeochemical Reactions. *Geochem. Geophys. Geosystems* **2023**, *24*, e2023GC010978. [CrossRef]
19. Liu, Z.; Gu, X.; Lian, M.; Wang, J.; Xin, M.; Wang, B.; Ouyang, W.; He, M.; Liu, X.; Lin, C. Occurrence, Geochemical Characteristics, Enrichment, and Ecological Risks of Rare Earth Elements in Sediments of "the Yellow River–Estuary–bay" System. *Environ. Pollut.* **2023**, *319*, 121025. [CrossRef]
20. Duman, M.; Kucuksezgin, F.; Eronat, A.H.; Talas, E.; İlhan, T.; Aydın, Ş. Combining Single and Complex Indices of Pollution with Grain Size Trend Analysis of Surficial Sediments in Edremit Gulf, Western Turkey. *Environ. Sci. Pollut. Res.* **2022**, *29*, 55609–55629. [CrossRef]
21. Sachithanandam, V.; Parthasarathy, P.; Sai Elangovan, S.; Kasilingam, K.; Dhivya, P.; Mageswaran, T.; Mohan, P.M. A Baseline Study on Trace Metals Concentration and Its Ecological Risk Assessment from the Coast of South Andaman Island, India. *Reg. Stud. Mar. Sci.* **2020**, *36*, 101242. [CrossRef]
22. Thabet, W.M.; Moneer, A.A.; Abdelwahab, O.; Ahdy, H.H.H.; Khedawy, M.; Shabaan, N.A. Ecological Risk Assessment of Metal Pollution in the Surface Sediments of Delta Region, Egypt. *Environ. Monit. Assess.* **2024**, *196*, 351. [CrossRef] [PubMed]
23. Lu, J.; Li, A.; Dong, J.; Pei, W.; Feng, X.; Wang, H. The Effect of Typhoon Talim on the Distribution of Heavy Metals on the Inner Shelf of the East China Sea. *Cont. Shelf Res.* **2021**, *229*, 104547. [CrossRef]
24. Szczuciński, W.; Niedzielski, P.; Rachlewicz, G.; Sobczyński, T.; Zioła, A.; Kowalski, A.; Lorenc, S.; Siepak, J. Contamination of Tsunami Sediments in a Coastal Zone Inundated by the 26 December 2004 Tsunami in Thailand. *Environ. Geol.* **2005**, *49*, 321–331. [CrossRef]
25. Rakib, M.R.J.; Rahman, M.A.; Onyena, A.P.; Kumar, R.; Sarker, A.; Hossain, M.B.; Islam, A.R.M.T.; Islam, M.S.; Rahman, M.M.; Jolly, Y.N.; et al. A Comprehensive Review of Heavy Metal Pollution in the Coastal Areas of Bangladesh: Abundance, Bioaccumulation, Health Implications, and Challenges. *Environ. Sci. Pollut. Res.* **2022**, *29*, 67532–67558. [CrossRef]
26. Hatje, V.; Pedreira, R.M.A.; de Rezende, C.E.; Schettini, C.A.F.; de Souza, G.C.; Marin, D.C.; Hackspacher, P.C. The Environmental Impacts of One of the Largest Tailing Dam Failures Worldwide. *Sci. Rep.* **2017**, *7*, 10706. [CrossRef]
27. Onyena, A.P.; Nkwoji, J.A.; Chukwu, L.O.; Walker, T.R.; Sam, K. Risk Assessment of Sediment PAH, BTEX, and Emerging Contaminants in Chanomi Creek Niger Delta, Nigeria. *Environ. Monit. Assess.* **2023**, *195*, 1080. [CrossRef]
28. Cundy, A.B.; Rowlands, F.M.; Lu, G.; Wang, W.-X. A Systematic Review of Emerging Contaminants in the Greater Bay Area (GBA), China: Current Baselines, Knowledge Gaps, and Research and Management Priorities. *Environ. Sci. Policy* **2022**, *131*, 196–208. [CrossRef]
29. Piskareva, V.M.; Gennadiev, A.N.; Lychagin, M.Y. The fluxes of polyarenes in the Don, Kuban and Volga river deltas. *Lomonosov Geogr. J.* **2021**, *5*, 14–23.
30. Caballero-Gallardo, K.; Olivero-Verbel, J.; Corada-Fernández, C.; Lara-Martín, P.A.; Juan-García, A. Emerging Contaminants and Priority Substances in Marine Sediments from Cartagena Bay and the Grand Marsh of Santa Marta (Ramsar Site), Colombia. *Environ. Monit. Assess.* **2021**, *193*, 596. [CrossRef]

31. Pavlenko, L.F.; Korablina, I.V.; Barabashin, T.O.; Ekilik, V.S. Priority Toxicants in Elements of Lower Don Ecosystem. *Water Resour* **2022**, *49*, 440–447. [CrossRef]
32. Strady, E.; Dang, T.H.; Dao, T.D.; Dinh, H.N.; Do, T.T.D.; Duong, T.N.; Duong, T.T.; Hoang, D.A.; Kieu-Le, T.C.; Le, T.P.Q.; et al. Baseline Assessment of Microplastic Concentrations in Marine and Freshwater Environments of a Developing Southeast Asian Country, Viet Nam. *Mar. Pollut. Bull.* **2021**, *162*, 111870. [CrossRef] [PubMed]
33. Zaki, M.R.M.; Ying, P.X.; Zainuddin, A.H.; Razak, M.R.; Aris, A.Z. Occurrence, Abundance, and Distribution of Microplastics Pollution: An Evidence in Surface Tropical Water of Klang River Estuary, Malaysia. *Environ. Geochem. Health* **2021**, *43*, 3733–3748 [CrossRef] [PubMed]
34. Bagaev, A.; Esiukova, E.; Litvinyuk, D.; Chubarenko, I.; Veerasingam, S.; Venkatachalapathy, R.; Verzhevskaya, L. Investigations of Plastic Contamination of Seawater, Marine and Coastal Sediments in the Russian Seas: A Review. *Environ. Sci. Pollut. Res* **2021**, *28*, 32264–32281. [CrossRef]
35. Apitz, S.E.; Agius, S. Anatomy of a Decision: Potential Regulatory Outcomes from Changes to Chemistry Protocols in the Canadian Disposal at Sea Program. *Mar. Pollut. Bull.* **2013**, *69*, 76–90. [CrossRef]
36. Mîndrescu, M.; Haliuc, A.; Zhang, W.; Carozza, L.; Carozza, J.-M.; Groparu, T.; Valette, P.; Sun, Q.; Nian, X.; Gradinaru, I. A 600 Years Sediment Record of Heavy Metal Pollution History in the Danube Delta. *Sci. Total Environ.* **2022**, *823*, 153702. [CrossRef]
37. Birch, G.F. A Review and Critical Assessment of Sedimentary Metal Indices Used in Determining the Magnitude of Anthropogenic Change in Coastal Environments. *Sci. Total Environ.* **2023**, *854*, 158129. [CrossRef]
38. Damasceno, F.L.; Alves Martins, M.V.; Senez-Mello, T.M.; Santos, L.G.C.; Mendonça Filho, J.G.; Pereira, E.; Figueira, R.; do Nascimento, C.A.; Arruda, S.; Castelo, W.F.L.; et al. Potential Ecological Risk by Metals in Sepetiba Bay (SE Brazil): Exporting Metals to the Oceanic Region. *J. S. Am. Earth Sci.* **2024**, *141*, 104934. [CrossRef]
39. Rudnick, R.L.; Gao, S. 4.1—Composition of the Continental Crust. In *Treatise on Geochemistry*, 2nd ed.; Holland, H.D., Turekian, K.K., Eds.; Elsevier: Oxford, UK, 2014; pp. 1–51. ISBN 978-0-08-098300-4.
40. Jaskuła, J.; Sojka, M.; Fiedler, M.; Wróżyński, R. Analysis of Spatial Variability of River Bottom Sediment Pollution with Heavy Metals and Assessment of Potential Ecological Hazard for the Warta River, Poland. *Minerals* **2021**, *11*, 327. [CrossRef]
41. El Ouaty, O.; El M'rini, A.; Nachite, D.; Marrocchino, E.; Rodella, I. Sediment Quality Indices for the Assessment of Heavy Metal Risk in Nador Lagoon Sediments (Morocco) Using Multistatistical Approaches. *Sustainability* **2024**, *16*, 1921. [CrossRef]
42. Hakanson, L. An Ecological Risk Index for Aquatic Pollution Control.a Sedimentological Approach. *Water Res.* **1980**, *14*, 975–1001. [CrossRef]
43. Birch, G.F.; Lee, J.-H.; Tanner, E.; Fortune, J.; Munksgaard, N.; Whitehead, J.; Coughanowr, C.; Agius, J.; Chrispijn, J.; Taylor, U.; et al. Sediment Metal Enrichment and Ecological Risk Assessment of Ten Ports and Estuaries in the World Harbours Project. *Mar. Pollut. Bull.* **2020**, *155*, 111129. [CrossRef] [PubMed]
44. Konstantinova, E.; Minkina, T.; Nevidomskaya, D.; Mandzhieva, S.; Bauer, T.; Zamulina, I.; Burachevskaya, M.; Sushkova, S. Exchangeable Form of Potentially Toxic Elements in Floodplain Soils along the River-Marine Systems of Southern Russia. *Eurasian J. Soil Sci.* **2021**, *10*, 132–141. [CrossRef]
45. Konstantinova, E.; Minkina, T.; Nevidomskaya, D.; Mandzhieva, S.; Bauer, T.; Zamulina, I.; Voloshina, M.; Lobzenko, I.; Maksimov, A.; Sushkova, S. Potentially Toxic Elements in Surface Soils of the Lower Don Floodplain and the Taganrog Bay Coast: Sources, Spatial Distribution and Pollution Assessment. *Environ. Geochem. Health* **2023**, *45*, 101–119. [CrossRef]
46. Konstantinova, E.; Minkina, T.; Mandzhieva, S.; Nevidomskaya, D.; Bauer, T.; Zamulina, I.; Sushkova, S.; Lychagin, M.; Rajput, V.D.; Wong, M.H. Ecological and Human Health Risks of Metal–PAH Combined Pollution in Riverine and Coastal Soils of Southern Russia. *Water* **2023**, *15*, 234. [CrossRef]
47. Minkina, T.M.; Fedorov, Y.A.; Nevidomskaya, D.G.; Pol'shina, T.N.; Mandzhieva, S.S.; Chaplygin, V.A. Heavy Metals in Soils and Plants of the Don River Estuary and the Taganrog Bay Coast. *Eurasian Soil Sci.* **2017**, *50*, 1033–1047. [CrossRef]
48. Minkina, T.M.; Nevidomskaya, D.G.; Pol'shina, T.N.; Fedorov, Y.A.; Mandzhieva, S.S.; Chaplygin, V.A.; Bauer, T.V.; Burachevskaya, M.V. Heavy Metals in the Soil–Plant System of the Don River Estuarine Region and the Taganrog Bay Coast. *J. Soils Sediments* **2017**, *17*, 1474–1491. [CrossRef]
49. Minkina, T.; Fedorenko, G.; Nevidomskaya, D.; Konstantinova, E.; Pol'shina, T.; Fedorenko, A.; Chaplygin, V.; Mandzhieva, S.; Dudnikova, T.; Hassan, T. The Morphological and Functional Organization of Cattails Typha Laxmannii Lepech. And Typha Australis Schum. and Thonn. under Soil Pollution by Potentially Toxic Elements. *Water* **2021**, *13*, 227. [CrossRef]
50. Mikhailenko, A.V.; Fedorov, Y.A.; Dotsenko, I.V. *Heavy Metals in the Components of the Landscape of the Sea of Azov*; Southern Federal University: Rostov-on-Don, Russia, 2018.
51. Matishov, G.G.; Bufetova, M.V.; Egorov, V.N. The Regulation of Flows of Heavy Metals Into the Sea of Azov According to the Intensity of Sedimentation of Water Self-Purification. *Sci. South Russ.* **2017**, *13*, 44–58. [CrossRef]
52. Sheverdyaev, I.V.; Kleschenkov, A.V. Revealing the Surge Phenomena Contribution of the Heavy Metals Inflow to the River Don Delta. *Morskoy Gidrofiz. Zhurnal* **2020**, *36*, 582–594. [CrossRef]
53. Tkachenko, A.N.; Tkachenko, O.V.; Lychagin, M.Y.; Kasimov, N.S. Heavy Metal Flows in Aquatic Systems of the Don and Kuban River Deltas. *Dokl. Earth Sci.* **2017**, *474*, 587–590. [CrossRef]
54. Tkachenko, O.V.; Tkachenko, A.N.; Lychagin, M.Y. Content of Heavy Metals in Water Objects of the Delta of Don: Seasonal and Spa-Tial Dynamics. *Geol. Geogr. Glob. Energy* **2016**, *2*, 76–84.

55. Sazykina, M.; Barabashin, T.; Konstantinova, E.; Al-Rammahi, A.A.K.; Pavlenko, L.; Khmelevtsova, L.; Karchava, S.; Klimova, M.; Mkhitaryan, I.; Khammami, M.; et al. Non-Corresponding Contaminants in Marine Surface Sediments as a Factor of ARGs Spread in the Sea of Azov. *Mar. Pollut. Bull.* **2022**, *184*, 114196. [CrossRef] [PubMed]
56. Bufetova, M.V. Estimation of Heavy Metal Fluxes (Pb, Cd) in the Abiotic Components of the Azov Sea Ecosystem. *Reg. Environ. Issues* **2018**, *4*, 70–73. [CrossRef]
57. Bufetova, M.V.; Fen, O.N. Assessment of Pollution of Azov Sea Bottom Sediments with Heavy Metals. *Proc. High. Educ. Establ. Geol. Explor.* **2016**, *3*, 45–51.
58. Mikhaylenko, A.V. Assessment of Mercury Content in Soils and Sediments of the Don River Delta. *Eng. J. Don* **2015**, *3*, 170.
59. Shishkin, A.N. (Ed.) *Environmental Atlas of the Black and Azov Seas*; Foundation "National Intellectual Resource": Moscow, Russia, 2019.
60. Venevsky, S.; Berdnikov, S.; Day, J.W.; Sorokina, V.; Gong, P.; Kleshchenkov, A.; Kulygin, V.; Li, W.; Misirov, S.; Sheverdyaev, I.; et al. Don River Delta Hydrological and Geomorphological Transformation under Anthropogenic and Natural Forcings. Dynamics of Delta Shoreline, Risk of Coastal Flooding and Related Management Options. *Ocean Coast. Manag.* **2024**, *258*, 107364. [CrossRef]
61. Tsygankova, A.E.; Bespalova, L.A. Intensity of Manifestation of Coastal Processes of the Sea of Azov for 1980–2020. *Ecol. Econ. Inform. Syst. Anal. Math. Model. Ecol. Econ. Syst.* **2022**, *1*, 132–138. [CrossRef]
62. Matishov, G.; Matishov, D.; Gargopa, Y.; Dashkevich, L.; Berdnikov, S.; Kulygin, V.; Arkhipova, O.; Chikin, A.; Shabas, I.; Baranova, O. *Climatic Atlas of the Sea of Azov*; U.S. Department of Commerce, National Oceanic and Atmospheric Administration, National Environmental Satellite, Data, and Information Service: Silver Spring, MD, USA, 2008.
63. Matishov, G.G.; Grigorenko, K.S. Causes of Salinization of the Gulf of Taganrog. *Dokl. Earth Sci.* **2017**, *477*, 1311–1315. [CrossRef]
64. Korotaev, V.N.; Chernov, A.V. Formation of Floodplains in the Lower Reaches of Large Plain Rivers under Base Level Fluctuations. *Vestn. Mosk. Univ. Ser. 5 Geogr.* **2018**, *4*, 29–38.
65. Vorobyova, L.A. *Theory and Practice of Chemical Analysis of Soils*; GEOS: Moscow, Russia, 2006.
66. Nikitin, B.A. Utochnenie k Metodike Opredeleniya Gumusa v Pochve. *Agrokhimiya* **1983**, *8*, 18–26.
67. Shaimukhametov, M.S. On the Method of Determination of Adsorbed Ca and Mg in Chernozemic Soils. *Pochvovedenie* **1993**, *12*, 105–111.
68. Gorbunov, N.I. Procedure of Sample Preparation to Mineralogical Analysis. In *Methods of Mineralogical and Micromorphological Study of Soils*; Gorbunov, N.I., Ed.; Nauka: Moscow, Russia, 1971; pp. 5–15.
69. *FR.1.31.2018.32143*; Measurement Technique M-049-PDO/18 "Methodology for Measuring the Mass Fraction of Metals and Metal Oxides in Powder Samples of Soils and Bottom Sediments Using the X-ray Fluorescence Method". Spectron NPO Ltd.: Saint Petersburg, Russia, 2018.
70. Badawy, W.M.; Duliu, O.G.; El Samman, H.; El-Taher, A.; Frontasyeva, M.V. A Review of Major and Trace Elements in Nile River and Western Red Sea Sediments: An Approach of Geochemistry, Pollution, and Associated Hazards. *Appl. Radiat. Isot.* **2021**, *170*, 109595. [CrossRef] [PubMed]
71. Abrahim, G.M.S.; Parker, R.J. Assessment of Heavy Metal Enrichment Factors and the Degree of Contamination in Marine Sediments from Tamaki Estuary, Auckland, New Zealand. *Environ. Monit. Assess.* **2008**, *136*, 227–238. [CrossRef] [PubMed]
72. Marchellina, A.; Soegianto, A.; Putranto, T.W.C.; Payus, C.M.; Irnidayanti, Y. Spatial Distribution and Pollution Assessment of Metals in Sediments along the Industrialized Coast of East Java, Indonesia. *Environ. Geochem. Health* **2024**, *46*, 205. [CrossRef]
73. Shabanov, M.V.; Marichev, M.S.; Minkina, T.M.; Rajput, V.D.; Bauer, T.V. Change in Eco-Geochemistry of Bottom Sediments in Copper–Pyrite Mining Area: A Case-Study of Karabash Copper Deposits. *Min. Inf. Anal. Bull.* **2023**, *5*, 117–134. [CrossRef]
74. Müller, G. Schadstoffe in Sedimenten-Sedimente Als Schadstoffe. *Mitt. Osterr. Geol. Ges.* **1986**, *79*, 107–126.
75. Nemerow, N.L. *Stream, Lake, Estuary, and Ocean Pollution*; Van Nostrand Reinhold: New York, NY, USA, 1991.
76. Usero, J.; González-Regalado, E.; Gracia, I. Trace Metals in the Bivalve Mollusc Chamelea Gallina from the Atlantic Coast of Southern Spain. *Mar. Pollut. Bull.* **1996**, *32*, 305–310. [CrossRef]
77. Tang, D.; Chen, M.; Huang, X.; Zhang, G.; Zeng, L.; Zhang, G.; Wu, S.; Wang, Y. SRplot: A Free Online Platform for Data Visualization and Graphing. *PLoS ONE* **2023**, *18*, e0294236. [CrossRef]
78. Konstantinova, E.; Minkina, T.; Nevidomskaya, D.; Bauer, T.; Fedorov, Y.; Zamulina, I.; Mandzhieva, S.; Kravtsova, N.; Voloshina, M.; Dudnikova, T.; et al. Establishment of Regional Background for Heavy Metals in the Soils of the Lower Don and the Taganrog Bay Coast. In Proceedings of the E3S Web of Conferences, Hyderabad, India, 24–26 September 2021; Volume 265.
79. MacDonald, D.D.; Ingersoll, C.G.; Berger, T.A. Development and Evaluation of Consensus-Based Sediment Quality Guidelines for Freshwater Ecosystems. *Arch. Environ. Contam. Toxicol.* **2000**, *39*, 20–31. [CrossRef]
80. Buchman, M.F. *Screening Quick Reference Tables (SQuiRTs)*; Office of Response and Restoration Division, National Oceanic and Atmospheric Administration: Seattle, WA, USA, 2008.
81. Macdonald, D.D.; Carr, R.S.; Calder, F.D.; Long, E.R.; Ingersoll, C.G. Development and Evaluation of Sediment Quality Guidelines for Florida Coastal Waters. *Ecotoxicology* **1996**, *5*, 253–278. [CrossRef] [PubMed]

minerals

MDPI

Article

High-Accuracy Image Segmentation Based on Hybrid Attention Mechanism for Sandstone Analysis

Lanfang Dong [1,2,3,4,*], Hao Gui [3], Xiaolu Yu [2], Xinming Zhang [4] and Mingyang Xu [5]

1 State Key Laboratory of Shale Oil and Gas Enrichment Mechanisms and Effective Development, Petroleum Exploration and Production Research Institute, China Petrochemical Corporation, Beijing 102206, China
2 Sinopec Key Laboratory of Petroleum Accumulation Mechanisms, Petroleum Exploration and Production Research Institute, China Petrochemical Corporation, Wuxi 214126, China; yuxl.syky@sinopec.com
3 School of Computer Science and Technology, University of Science and Technology of China, Hefei 230027, China; guihao@mail.ustc.edu.cn
4 Institute of Advanced Technology, University of Science and Technology of China, Hefei 230031, China; xm_zhang@mail.ustc.edu.cn
5 Anhui Rank Artificial Intelligent Technology Co., Ltd., Hefei 230088, China; xumingyang@ranktek.cn
* Correspondence: lfdong@ustc.edu.cn

Abstract: Mineral image segmentation based on computer vision is vital to realize automatic mineral analysis. However, current image segmentation methods still cannot effectively solve the problem of sandstone grains that are adjoined and concealed by leaching processes, and the segmentation performance of small and irregular grains still needs to be improved. This investigation explores and designs a Mask R-CNN-based sandstone image segmentation model, including a hybrid attention mechanism, loss function construction, and receptive field enlargement. Simultaneously, we propose a high-quality sandstone dataset with abundant labels named SMISD to facilitate comprehensive training of the model. The experimental results show that the proposed segmentation model has excellent segmentation performance, effectively solving adhesion and overlap between adjacent grains without affecting the classification accuracy. The model has comparable performance to other models on the COCO dataset, and performs better on SMISD than others.

Keywords: mineral image analysis; sandstone grain segmentation; deep learning; Mask R-CNN

1. Introduction

Most mineral identification and textural characterization in coarse-grained clastic sediments (such as sandstones) necessitate the use of microscopy assessment, which involves interpretation and the gathering of counting statistics using a light microscope [1–6]. The analysis of sandstone images is a vital part of geological exploration research, as rocks are rich in hydrocarbon resources [7]. Sedimentary petrology is a specialized field that demands significant expertise to interpret and classify porosity, minerals, matrix materials, and mineral cements, along with their associated subclasses. A range of objective lenses, optical techniques, and light sources are employed to identify and quantify components. Certain components in sandstone can be small, dark, or opaque, posing challenges for interpretation even to an experienced petrologist. In such cases, methods beyond light microscopy, such as scanning electron microscopy or bulk measurement techniques like X-ray diffraction, are often utilized to assist in mineral identification [8]. Traditionally, the analysis of sandstone thin sections is primarily conducted by professionals, which is labor-intensive, expensive, and subjective.

At present, classic mineral image segmentation technology primarily relies on the low-level visual information of image pixels and is mainly categorized into three types: (1) the threshold-based mineral image segmentation algorithm, which compares each pixel in the input images with a preset threshold value to segment the target areas [9–12]; (2) the region-based mineral image segmentation algorithm that divides the original image into different pixel regions, separating the target areas from the background [13–15]; and (3) the specific theory-based mineral image segmentation algorithm that employs more targeted computational methods such as cluster analysis to separate mineral grain images [16]. Although these classic mineral image segmentation methods have improved the efficiency of sandstone analysis, they cannot effectively address the issue of adhesion and overlap between adjacent grains, and their segmentation performance for small and irregular grains is relatively poor. Additionally, the classic image segmentation algorithm requires manual tuning for different types of mineral grains, which lowers the efficiency of mineral image segmentation and increases the time consumption.

Owing to the booming development of deep learning and the exceptional feature extraction capabilities of CNNs (convolutional neural networks), deep learning-based approaches have been increasingly employed in mineral image segmentation. These methods, diverging from traditional ones, adhere to an end-to-end paradigm and significantly surpass conventional methods in accuracy and efficiency. Furthermore, they are data-driven methods (i.e., the larger the dataset or the more refined the data analysis, the higher the achievable accuracy), allowing for enhanced performance with continuous data expansion. Some studies have explored and applied it in mineral image segmentation tasks. An RDU (R: residual connection; DU: DUNet) ore image segmentation model was proposed to estimate the grain size of ore fragments in conveyor belts, which can adjust the receptive field adaptively according to the size and shape of different ore fragments and achieve accurate segmentation [17]. Deep learning-based methods for mineral image segmentation also excel in segmenting adherent, overlapping, and multi-scale mineral grains, effectively addressing the typical challenges encountered in previous approaches [18]. However, existing image segmentation methods mainly focus on semantic segmentation, which cannot meet the requirements to compute specified morphological data in the mineral analysis, such as circularity, particle size, and contact relationship. Furthermore, these models are mostly designed for scenarios with clean image backgrounds, whereas in sandstone analysis scenarios, the image backgrounds are complex and filled with indistinguishable fillers, which makes it a challenge to segment grains out [19].

In order to solve the above problems and further improve the application prospects of automatic sandstone analysis, we propose a high-accuracy instance segmentation model based on Mask R-CNN [20] by introducing a hybrid attention mechanism to better adapt to the complex shapes of sandstone particles. Secondly, we design a shape-aware training loss function for the improved model and conduct an ablation experiment to confirm its advantages. Additionally, original convolution is replaced with dilated convolution for the purpose of obtaining more global information. Finally, this experiment establishes a dataset of sandstone instance segmentation with 40,122 grain labels.

2. Methodology

Section 2 mainly introduces various methods and ideas involved in the improved model of the sandstone image segmentation system. In the model-building stage, backbone selection, module setting, and loss function design are the main considered aspects to improve the segmentation model performance and solve edge blurring and incomplete segmentation in the segmentation process.

2.1. Mask R-CNN

The deep learning-based Mask R-CNN network segmentation method has superior performance and can achieve good results in the sandstone microscopic image segmentation task [21]. We customize and optimize the Mask R-CNN network to form a task specified model, based on the distribution characteristics of sandstone grains in sandstone microscopic images.

The Mask R-CNN network is divided into three steps: object localization, object category calculation, and segmentation mask prediction. The process of the Mask R-CNN network is as follows: the image passes through the ResNet backbone network, and different levels of feature maps are obtained using the feature pyramid. This facilitates the model in extracting features at different levels. It then enters the region proposal network to generate candidate areas where grains might be present. On the one hand, a classifier determines whether pixels belong to sandstone grains or the background. On the other hand, a box regressor corrects the boundaries of the sandstone grains. The combination of these two steps forms the candidate target areas. Finally, through a fully convolutional neural network, accurate segmentation results are achieved, completely extracting the sandstone grains. The processing procedure is shown in Figure 1.

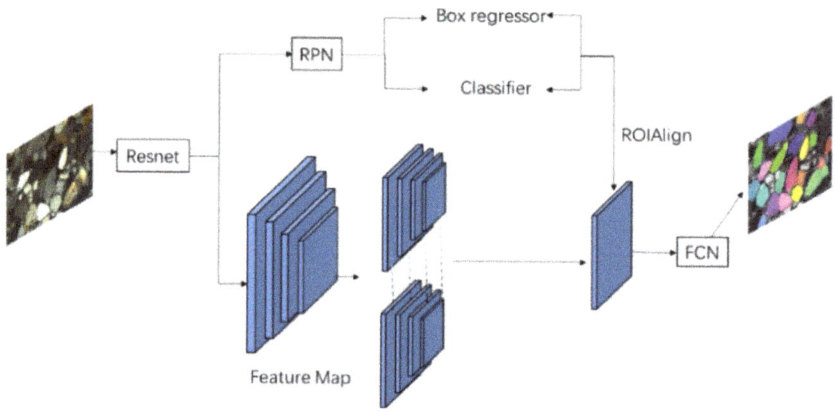

Figure 1. Original Mask R-CNN sandstone image segmentation processing flowchart.

2.2. SE-Net

Squeeze-and-Excitation Network (SE-Net [22]) introduces the concept of channel attention. By modeling and assigning weights to feature channels, it forms the parameters that can be learned and updated and continuously increases the weights of useful feature channels to optimize the generalization capability of the model. In this paper, SE-Net is added to the backbone recognition network to improve the spatial information processing capability of the model and therefore achieve better results for sandstone grains at different scales. The specific structure of SE-Net is shown in Figure 2.

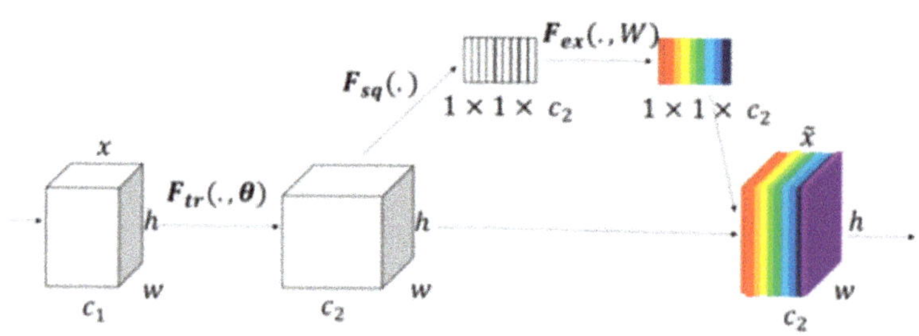

Figure 2. SE-Net structure diagram.

The first operation is F_{tr} conversion, which converts the input x with a feature channel number C_1 by a series of general transformations such as convolution to obtain a feature with a feature channel number C_2. The F_{tr} operation is shown in Equation (1).

$$F_{tr} : X \to U, \;\; X \in R^{H \times W \times c_1}, \;\; U \in R^{H \times W \times c_2} \tag{1}$$

The Squeeze operation is shown in Equation (2), where the two-dimensional feature channels are mapped by compression to obtain a real number, which is connected to form a one-dimensional vector to obtain the global distribution and construct the global receptive field of the model. The dimension number of the vector equals the number of input feature channels.

$$z_c = F_{sq}(u_c) = \frac{1}{H \times W}\sum_{i=1}^{H}\sum_{j=1}^{W} u_c(i,j) \tag{2}$$

Next is the Excitation operation. The parameter w is used to generate the weights for each feature channel. Multiply W_1 (which dimension is $C/r \times C$) and z to reduce the computational complexity by scale operation. The dimension of the feature map of W_1z is still $1 \times 1 \times C/r$; it is then passed through the ReLU layer and multiplied with W_2 (which dimension is $C \times C/r$) to obtain the feature map dimension $1 \times 1 \times C$. Finally, the Sigmoid is derived to generate the weights (s) of the feature maps (at a total of C). The parameter s incorporates the feature map information of each feature channel and is part of the neural network, which can be learned and optimized.

$$s = F_{ex}(z, W) = \sigma(g(z, W)) = \sigma(W_2\delta(W_1 z)) \tag{3}$$

Finally, the weights calculated by the model are weighted onto the original feature map by the previous channels through the scale operation, as shown:

$$\widetilde{x_c} = F_{scale}(u_c, s_c) = s_c u_c \tag{4}$$

2.3. Coordinate Attention and Spatial Attention

The channel attention mechanism models global information through channels, allowing the model to effectively extract sandstone grains. When extracting, the location information of the sandstone is directly related to the accuracy of the boundary fit. In this paper, a CA + SP hybrid attention mechanism is proposed to optimize the model by combining the coordinate attention mechanism [23] and the spatial attention mechanism [24].

Coordinate attention takes a similar operation to channel attention in both horizontal and vertical directions to obtain relatively independent feature maps in both directions, effectively preserving one-dimensional location information and establishing spatial long-range dependence in one dimension. This mechanism is very sensitive to coordinate information and can effectively pinpoint spatial coordinate information. Similar to the channel attention mechanism, the coordinate attention mechanism first performs coordinate position encoding and then generates coordinate attention. A coordinate attention module can be seen as a computational unit used to augment feature representation capabilities. It can take any intermediate tensor $X = [x_1, x_2, \ldots, x_C] \in R^{C \times H \times W}$ as input and produces an output $Y = [y_1, y_2, \ldots, y_C]$ of the same size with enhanced representational power. The horizontal and vertical directions are treated separately and computed using pooling kernels of dimensions (H, 1) and (1, W), giving the following outputs for the vertical and horizontal channels.

$$z_c^h(h) = \frac{1}{W}\sum_{0 \le i < W} x_c(h, i),\ z_c^w(w) = \frac{1}{W}\sum_{0 \le j < W} x_c(j, w) \tag{5}$$

The two feature maps generated by the previous module are cascaded and then transformed using a shared 1×1 convolution to perform the transformation F_1, expressed as in Equation (5); the generated $f \in R^{C/r \times (H+W)}$ is the intermediate feature map during the computation, where r denotes the down-sampled ratio, which is used to control the size of the module, like the SE module.

$$f = \delta\left(F_1\left(\left[z^h, z^w\right]\right)\right) \tag{6}$$

Next, f is cut into two direction-independent tensors, $f^h \in R^{\frac{C}{r} \times H}, f^w \in R^{\frac{C}{r} \times W}$, and then using the two 1×1 convolutions F_h and F_w, we transform two tensors to the same number of channels as the input and output, as in the following equation:

$$g^h = \sigma\left(F_h\left(f^h\right)\right), g^w = \sigma\left(F_h\left(f^h\right)\right) \tag{7}$$

The two are then expanded as attention weights, and final output of the CA module is

$$\widetilde{y_c}(i,j) = x_c(i,j) \times g_c^h(i) \times g_c^w(j) \tag{8}$$

The spatial attention mechanism can simulate the function of the human eye and extract the parts of interest to the model, which generates a mask for the space, and draws out one way, which undergoes similar operations as described above to form the spatial attention mechanism, effectively extracting its relative spatial information.

2.4. Dilated Convolution

The up-sampling process of bilinear interpolation has large errors, leading to problems such as distortion when generating sandstone grain contours, and this paper achieves refinement of sandstone contours by introducing dilated convolution. The dilated convolution [25] has the following main functions:

1. Expanding the Mask R-CNN network receptive fields more efficiently while taking into account image resolution;
2. Changing the size of the convolution kernel and the perceptual field of the model by adjusting the expansion rate (r) to obtain multi-scale global semantic information.

The dilated convolution is shown in Figure 3. Only non-zero elements play a role in the calculation, and the dilated convolution fills the ordinary convolution with zeros to increase the size of the receptive field, as shown in Equation (9).

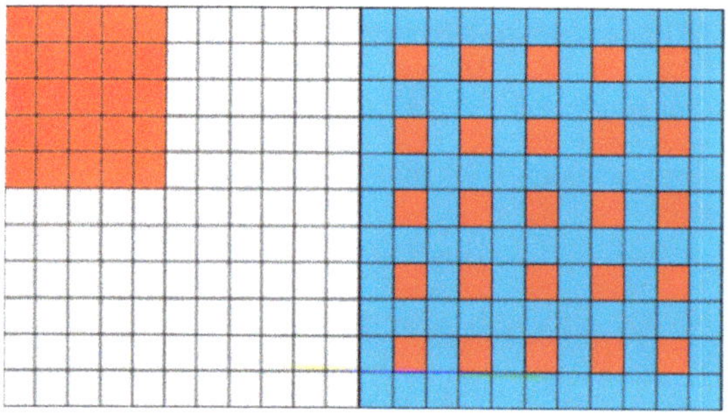

Figure 3. Dilated convolution diagram.

$$K = k + (k-1)(r-1) \tag{9}$$

K is the size of the expanded convolution kernel, k is the size of the original convolution kernel, and r represents the expansion rate. As shown in the figure, when k = 5 and r = 2, compared with the ordinary 5×5 convolution on the left, the dilated convolution expands the convolution kernel to 11×11, and the range of the sensory field is greatly improved.

3. Methods and Materials

3.1. Experimental Methods

3.1.1. Hybrid Attention Mechanism

In this paper, the SE module is embedded in the backbone network ResNet and its processing flow is shown in Figure 4. In contrast to the SE-Net structure mentioned earlier, Squeeze takes global average pooling and Excitation takes two full connection layers for computation. To maintain the ResNet structure, it is also necessary to ensure that the input and output dimensions are the same: the feature dimension is first reduced to 1/4 of the input, and then the dimension is recovered after ReLU activation. This operation not only complies more with the specification of the Excitation operation but also reduces the computational costs by reducing the dimensionality. The neural network architecture with two fully connected layers allows for better modeling of correlations between channels. Finally, analogous to the SE-Net model, the scale operation is set to weight the weights by channel onto the corresponding original feature maps to form the SE-ResNet model.

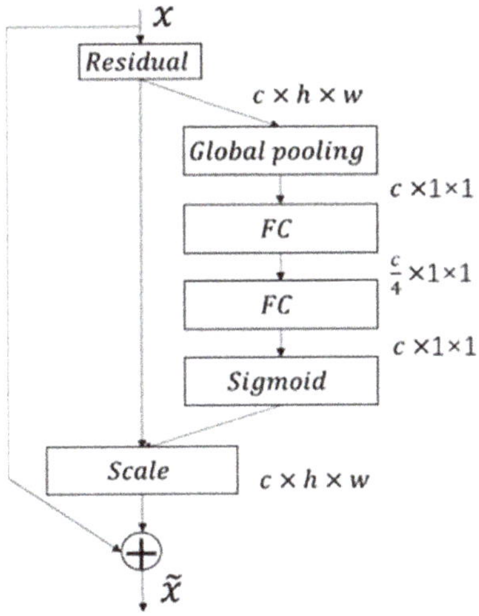

Figure 4. SE-ResNet module architecture diagram.

Additionally, a CA + SP-ResNet backbone network is proposed to optimize the model by combining the coordinate attention mechanism and the spatial attention mechanism in a re-weighted way, which is shown in Figure 5.

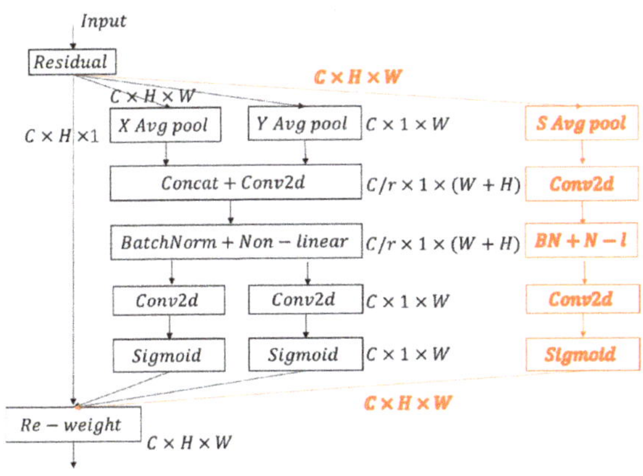

Figure 5. Coordinate attention + spatial attention ResNet module architecture diagram.

After the spatial attention mechanism, the output section of the module should also multiply the coordinate attention output above by the effect of spatial attention and modify the final output to Equation (10).

$$y(i,j) = \tilde{y}(i,j) \times g_c^s(i,j) \tag{10}$$

Finally, the output of SE-ResNet and CA + SP-ResNet also uses the similar re-weighted method to obtain the final result.

3.1.2. Loss Function

Sandstone grains are often dense and complex in shape. To reduce segmentation error, an accurate sandstone grain boundary is of great importance. In fact, it is necessary to compare the real grain area with the predictions to ensure that their boundaries and contours fit perfectly. Thus, we use DiceLoss [26] to calculate the segmentation error, which is used to calculate the Dice (the graphical similarity between the true and predicted segmentation results of the sandstone grains). Let the true result of segmentation be $|T|$ and the predicted result be $|P|$, and then $|T \cap P|$ be the dot product of the binary plot of the predicted segmentation result and the true segmentation result. The elements in $|T|$ and $|P|$ directly take the summation. The similarity coefficient Dice and segmentation error DiceLoss are as follows:

$$\mathrm{Dice}(P,T) = \frac{|P \cap T|}{(|P+T|)/2} = \frac{2 * \mathrm{gtmask} \cap \mathrm{predmask}}{\mathrm{gtmask} + \mathrm{predmask}} \tag{11}$$

$$\mathrm{DiceLoss}(P,T) = 1 - 2\frac{|P \cap T|}{|P|+|T|} \tag{12}$$

To avoid the denominator of the equation being zero and to reduce over-fitting, smoothing is added to the numerator and denominator at the same time:

$$\mathrm{DiceLoss}(P,T) = 1 - 2\frac{|P \cap T| + \mathrm{smooth}}{|P| + |T| + \mathrm{smooth}} \tag{13}$$

Since cross-entropy is only a form of proxy, we directly adopt DiceLoss as the loss function. The real target of the segmentation is to maximize the overlap between the predicted results and the true segmentation results, i.e., the Dice coefficient, thus minimizing DiceLoss. Also, the prediction of segmentation network accuracy performance often needs to be tested with IoU, which is very similar to DiceLoss; this operation allows the performance metric to be trained and optimized directly as a component of the loss function. Due to the non-convexity of the DiceLoss function, there can be problems such as gradient explosion, which typically leads to problems such as more unstable training and difficulty in converging the training. Through several experiments, this paper uses Log-Cosh DiceLoss, based on the Log-Cosh function, to smooth the DiceLoss. The Cosh function as well as the Log-Cosh DiceLoss equation, i.e., $L_{lc\text{-}dce}$, are shown in Equations (14) and (15).

$$\mathrm{Cosh}\, x = \frac{e^x + e^{-x}}{2} \tag{14}$$

$$L_{lc-dce} = \log(\cosh(\mathrm{DiceLoss})) \tag{15}$$

By changing the segmentation loss from L_{mask} to $L_{lc\text{-}dce}$, we obtain a more refined simulation of sandstone microscopic image boundaries, effectively improving the segmentation effect and contour refinement for irregular sandstone grains.

3.2. Material Preparation

Physical thin sections were acquired in a standard manner, in which blue epoxy (variable hue) was impregnated into the sandstone (a strictly coarse-grained clastic sediment), an approximately 30 μm thick thin section on a glass slide was prepared, and a mechanical polish was applied to the surface. Immediately prior to imaging, mineral oil and a cover slip were applied to the thin section. Generally, few label sets for petrology disciplines exist and their generation is extremely time-consuming. The images, used for labeling, were acquired on several different optical microscopes using objective lenses with 5× or 10× magnification.

The dataset SMISD has 288 typical labeled, single-polarized sandstone grain images with a cumulative total of 40,122 sandstone grains. An approach was taken to expand the dataset by rotating the images counterclockwise by different degrees, as shown in Figure 6, to a total of 1121 images with a total of 153,466 grains.

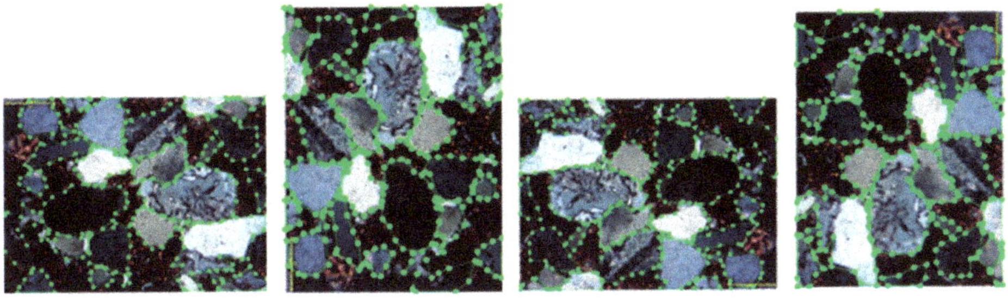

Figure 6. Image enhancement methods for sandstone microscopic image segmentation dataset.

In order to obtain better generalization ability, in this paper, 85% of them are divided into a training set and the remaining 15% are divided into a test set.

4. Results

4.1. Experiments and Analysis

To test the performance of the improved Mask R-CNN model based on the above theory for sandstone microscopic image segmentation, several experiments were conducted based on the traditional Mask R-CNN network and the improved Mask R-CNN network, and the experimental environment is shown in Table 1.

Table 1. Table of experimental environment.

Software/Hardware	Configuration
Operating system	Ubuntu 20.04
Memory	32 GB
CPU	Intel(R) Core(TM) i9-10920X CPU @ 3.50 GHz
GPU	NVIDIA GeForce RTX 3090
Related software	Python3.8/Torch1.8.0/cuda11.1

The basic parameters of the Mask R-CNN network were adjusted to obtain the best experimental results, and the hyper-parameters were set as follows: the model threshold was set to 0.5, the MINI_MASK was set to (28,28) (to facilitate the detection of fine grains), ResNet50 was used for the skeleton network (to increase the model processing speed), and the maximum number of grains present in a single image was set to 1000 (taking into account the dense distribution of grains in some sandstone micro-graphs).

4.2. Performance Metric

For the sandstone microscopic image segmentation task, the IoU is fixed and the average precision rate and average recall rate are taken as evaluation metrics.

4.2.1. Accuracy and Recall Rate, AP

To introduce the performance metric of sandstone microscopic image segmentation models (i.e., the generalization ability), the following four concepts are introduced: TP and FP, respectively, refer to the number of correct and incorrect determinations for all samples judged to be sandstone grains. TN and FN, respectively, refer to the number of correct and incorrect determinations for samples judged to be non-sandstone grains. Accuracy is the probability that all samples predicted to be sandstone grains are indeed sandstone grains, and it addresses the accuracy of the prediction results for a specific sample and sets the accuracy rate as P, given by the following equation:

$$P = \frac{TP}{TP + FP} \tag{16}$$

The recall rate is the probability that the model correctly predicts a sample that is indeed sandstone grains and it focuses on the prediction of the specific sample. If we set the recall as R, the equation is as follows:

$$R = \frac{TP}{TP + FN} \tag{17}$$

Accuracy and recall rate are interactive, with both being high when the model generalization is strong. The two are usually negatively correlated. If the X-axis is the recall rate and the Y-axis is the precision rate, the PR curve is obtained by tracing the points. The average accuracy (AP) is the PR curve integrated from 0 to 1. The calculation formula is as follows:

$$AP = \int_0^1 P(r)dr \tag{18}$$

The calculation is simplified by smoothing, which is calculated as follows:

$$AP = \frac{1}{11} \sum_{i=0,0.1,\ldots,1.0} \text{smooth}(i) \tag{19}$$

4.2.2. IoU

IoU [27] represents the overlap ratio between the generated and real image segmentation regions. The calculation of IoU is shown in Equation (20), where C is the real image segmentation area and G is the segmentation area generated by the model. A diagram of IoU is shown in Figure 7. The left side of the image is the intersection of the two and the right side is the merging of the two.

$$IoU = \frac{Area(C) \cap Area(G)}{Area(c) \cup Area(G)} \tag{20}$$

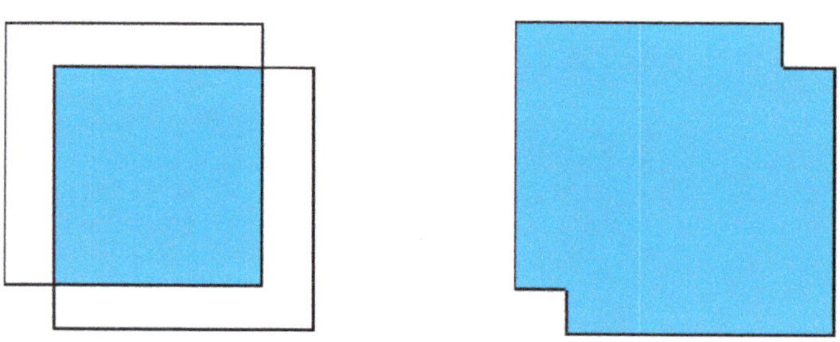

(**a**) Area(C) ∩ Area(G) (**b**) Area(c) ∪ Area(G)

Figure 7. Schematic diagram of the IoU.

The following six metrics for characterizing the performance of image segmentation can be obtained based on the above concepts, separately shown in Table 2.

Table 2. Image segmentation performance indicators.

	Average Precision (AP)
AP	AP When IoU = 0.50:0.05:0.95
AP_{50}	AP When IoU = 0.50
AP_{75}	AP When IoU = 0.75
	AP Across Scales
AP_S	AP When the grain is small: pixel area < 32^2
AP_M	AP When the grain is medium: 32^2 < pixel area < 96^2
AP_L	AP When the grain is large: pixel area > 96^2

AP_{50} means that the model is judged to be correct when the IoU of the predicted and real grain area is greater than 0.5; then, the classification AP of the model is calculated. AP_{75} needs a higher IoU threshold; thus, the segmentation needs to be more refined.

AP, i.e., AP:IoU = 0.50:0.05:0.95:, means, respectively, setting the correct determination IoU to 0.5, 0.55, 0.6, 0.65, 0.7, 0.75, 0.8, 0.85, 0.9, or 0.95 and then averaging it after calculating the AP value to obtain the final AP.

At the same time, the test results are classified according to the area occupied by the pixels. AP Across Scales means that the detection targets are classified according to area into three categories—small, medium, and large—and then the predictive ability of the model is judged.

The accuracy rate of the model was calculated separately by several indicators such as IoU accuracy and target size to obtain the final image segmentation performance metrics, as shown in Table 2.

Compared to the conventional model, the improved model identified significantly more grains, segmented grain contours were significantly finer, and AP values were improved by about 5%.

The effectiveness of the hybrid attention mechanism, its shape awareness, and its dilated convolution are shown in Tables 3–5, respectively.

Table 3. The effectiveness of the attention mechanism.

Backbone	**SMISD (%)**		
	$\mathbf{AP_S}$	$\mathbf{AP_M}$	$\mathbf{AP_L}$
ResNet	15.4	37.4	41.2
+SE	18.9	38.1	42.3
+X attention	19.3	38.9	41.9
+Y attention	19.2	38.7	42.0
+CA	20.3	39.5	43.1
+CA + SP	20.8	39.7	43.2

Table 4. The effectiveness of the shape-aware loss function.

Loss Function	**SMISD (%)**		
	AP	$\mathbf{AP_{50}}$	$\mathbf{AP_{75}}$
L_{mask}	33.2	38.6	29.1
L_{dice}	36.3	41.3	37.3

Table 5. The effectiveness of dilated convolution.

Type	**SMISD (%)**		
	AP	$\mathbf{AP_{50}}$	$\mathbf{AP_{75}}$
Original convolution	33.2	38.6	29.1
Dilated convolution	37.9	43.2	34.2

4.2.3. Results and Visualization

The sandstone micrographic images used to visualize the effectiveness our model are shown in Figure 8. The results of segmentation when directly using the original Mask R-CNN network are shown in Figure 9; the contour fitting was poor for sandstone grains of varying sizes and shapes. The network can only fit about 10% when grains are small.

(**a**) Large grain, plain light (**b**) Large grain, cross-polarized

(**c**) Small grain, plain light (**d**) Small grain, cross-polarized

Figure 8. Images used to visualize the effectiveness.

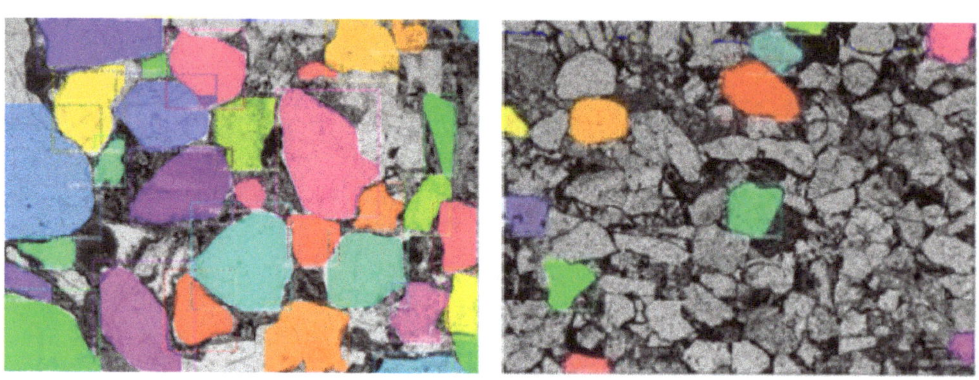

(**a**) Large grains (**b**) Small grains

Figure 9. Processing results of original Mask R-CNN network.

After the above modification, new segmented images are shown in Figure 10. Obviously, the model can identify about 80% of the grains regardless of their size, and the contours fit more closely. The mis-segmentation and under-segmentation are greatly reduced, and the generalization ability of the model is greatly improved.

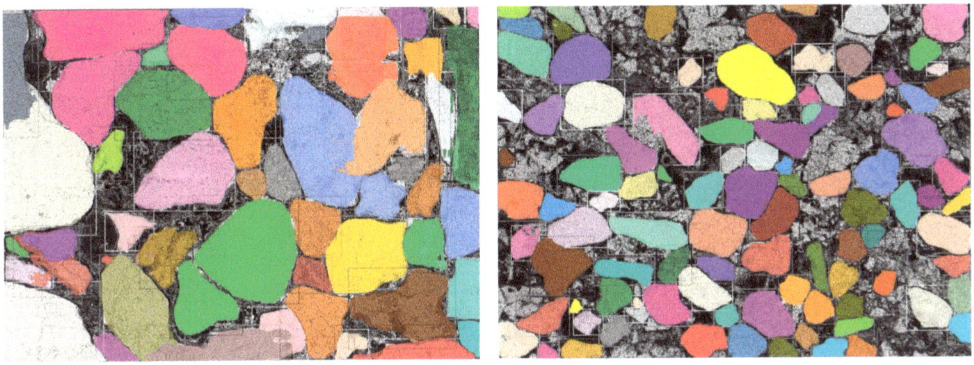

(**a**) Large grains (**b**) Small grains

Figure 10. Processing results of the improved Mask R-CNN network in this paper.

Comparison of the Results of the Improved Mask R-CNN Network in Fitting Irregular Sandstone Grain Images

For the irregular, narrow sandstone grains, the new model is far superior to the original model in segmentation integrity as well as fit, which is shown in Figure 11.

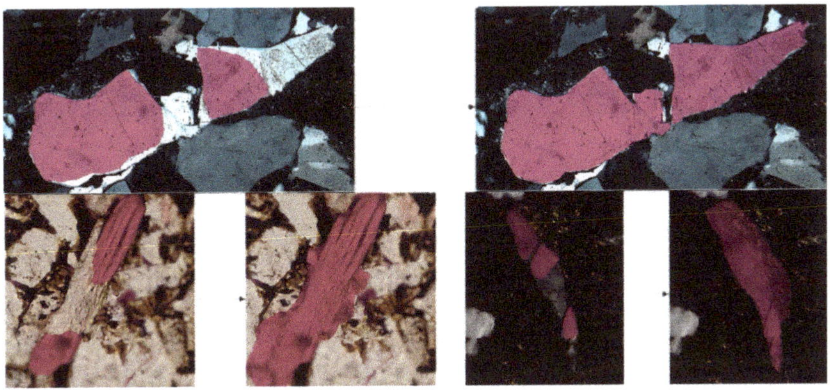

Figure 11. Comparison of the results of the original and improved Mask R-CNN in fitting irregular sandstone grain images.

Experiments on the Performance of the Improved Segmentation Network

The performance of the proposed image segmentation network is predicted by testing 238 sandstone images in the test set. Experiments were conducted using Mask R-CNN, HTC, PointRend, the latest Refine Mask, and the improved image segmentation algorithm (labeled Hybrid + Loss in the table), based on the publicly available COCO image segmentation dataset [28] and the sandstone microscopic dataset in this paper. For the best results, the recognition backbone networks all adopted ResNet101.

As shown in Table 6, our algorithm achieved the best results out of all the algorithms in the sandstone microscopic image segmentation task, based on SMISD. Compared to the conventional Mask R-CNN network, the improvement is 4.3%, 5.2%, and 6.9% for each of the three grain types (large, medium, and small), which is slightly better than the state-of-the-art Refine Mask network. Hybrid + Loss gained significant improvement in AP values for all sizes of grains, which indicates that the image segmentation algorithm tailored to the sandstone microscopic image segmentation task can effectively identify and extract sandstone grains, and can better fit the contours to achieve fine segmentation of sandstone grains.

Table 6. Performance comparison of segmentation algorithms.

Dataset / Algorithm	COCO (%)				SMISD (%)			
	AP	AP_S	AP_M	AP_L	AP	AP_S	AP_M	AP_L
Mask R-CNN	39.6	27.2	49.0	57.7	32.3	15.4	37.4	41.2
HTC	41.2	27.2	51.9	61.5	33.9	15.6	38.9	44.0
PointRend	41.1	27.8	52.0	62.0	35.1	16.3	39.9	45.7
RefineMask	41.8	28.6	53.1	62.8	36.7	18.0	41.1	47.3
Hybrid + Loss	41.7	28.9	52.7	62.5	37.9	19.7	42.6	48.1

5. Discussion

Channel attention can enhance the network's ability to extract image information. Coordinate attention can improve the model's ability to locate boundaries. Spatial attention can enhance the model's receptive capability and optimize the model's generalization ability, that is, its performance on images that have not been used for training. By introducing the hybrid attention mechanism, the model proposed in this paper surpasses other models on SMISD. With the goal of the task—to make the boundaries of grain segmentation results more precise—as the performance evaluation index, a shape-aware loss function can improve the model's segmentation effect on grain contours, especially irregular grains. Therefore, the model adopts Log-Cosh DiceLoss as the model's loss function. Finally, the dilated convolution can expand the receptive field and segment by combining more regional information surrounding the grains, which makes model perform better. Additionally, the dataset with rich labeled grains helps the model fully learn the segment rules, which is vital for data-driven methods.

6. Conclusions

Intelligent analysis of images of thin sandstone sections is of high research and application value, but the search for automatic segmentation and recognition is challenging because of the variety and complexity of sandstone images. This paper designed a high-accuracy sandstone segmentation model by mixing several attention mechanisms and applying appropriate loss functions and dilated convolution. The segmented particle images can be classified using other methods to obtain their mineralogical categories. The research on the automatic analysis of thin sandstone sections has made some progress in this paper, but it is difficult to identify filler, especially heterogeneous groups and cementation, and there is still much room for improvement in recognition accuracy.

Author Contributions: Methodology, H.G.; software, M.X.; validation, X.Z.; investigation, H.G. and L.D.; resources, X.Y.; writing—original draft preparation, H.G.; writing—review and editing, X.Z.; visualization, H.G.; supervision, L.D.; project administration, L.D.; funding acquisition, X.Y. All authors have read and agreed to the published version of the manuscript.

Funding: This research is funded by the SINOPEC Key Laboratory of Petroleum Accumulation Mechanisms's Microscopic Panoramic Segmentation of Dense Sandstone Open Fund, funding number 33550007-22-ZC0613-0038, and SINOPEC Excellent Youth Technology Innovation Fund, funding number P19028.

Data Availability Statement: Restrictions apply to the availability of these data. Data were obtained from the SINOPEC Key Laboratory of Petroleum Accumulation Mechanisms and are available from the author Xiaolu Yu with the permission of SINOPEC Key Laboratory of Petroleum Accumulation Mechanisms.

Acknowledgments: Thanks to the SINOPEC Key Laboratory of Petroleum Accumulation Mechanisms for the given sandstone microscopic images.

Conflicts of Interest: Xiaolu Yu is employee of Sinopec. The paper reflects the views of the scientists and not the company. Mingyang Xu are employee of Anhui Rank Artificial Intelligent Technology Co., Ltd. The paper reflects the views of the scientists and not the company.

References

1. Taylor, T.R.; Giles, M.R.; Hathon, L.A.; Diggs, T.N.; Braunsdorf, N.R.; Birbiglia, G.V.; Kittridge, M.G.; Macaulay, C.I.; Espejo, I.S. Sandstone diagenesis and reservoir quality prediction: Models, myths, and reality. *AAPG Bull.* **2010**, *94*, 1093–1132. [CrossRef]
2. Milliken, K. *Late Diagenesis and Mass Transfer in Sandstone Shale Sequences*; Elsevier: Amsterdam, The Netherlands, 2003; Volume 7.
3. Makowitz, A.; Milliken, K.L. Quantification of brittle deformation in burial compaction, Frio and Mount Simon Formation sandstones. *J. Sediment. Res.* **2003**, *73*, 1007–1021. [CrossRef]
4. Makowitz, A.; Lander, R.; Milliken, K. Diagenetic modeling to assess the relative timing of quartz cementation and brittle grain processes during compaction. *AAPG Bull.* **2006**, *90*, 873–885. [CrossRef]
5. Dutton, S.P.; Loucks, R.G.; Day-Stirrat, R.J. Impact of regional variation in detrital mineral composition on reservoir quality in deep to ultradeep lower Miocene sandstones, western Gulf of Mexico. *Mar. Pet. Geol.* **2012**, *35*, 139–153. [CrossRef]
6. Dutton, S.P.; Loucks, R.G. Diagenetic controls on evolution of porosity and permeability in lower Tertiary Wilcox sandstones from shallow to ultradeep (200–6700 m) burial, Gulf of Mexico Basin, USA. *Mar. Pet. Geol.* **2010**, *27*, 69–81. [CrossRef]
7. McRae, L.; Holtz, M.; Hentz, T. *Strategies for Reservoir Characterization and Identification of Incremental Recovery Opportunities in Mature Reservoirs in Frio Fluvial-Deltaic Sandstones, South Texas: An Example from Rincon Field, Starr County*; Topical report; U.S. Department of Energy: Washington, DC, USA, 1995. Available online: https://www.osti.gov/servlets/purl/123238 (accessed on 17 March 2024).
8. Saxena, N.; Day-Stirrat, R.J.; Hows, A.; Hofmann, R. Application of deep learning for semantic segmentation of sandstone thin sections. *Comput. Geosci.* **2021**, *152*, 104778. [CrossRef]
9. Perez, C.A.; Estévez, P.A.; Vera, P.A.; Castillo, L.E.; Aravena, C.M.; Schulz, D.A.; Medina, L.E. Ore grade estimation by feature selection and voting using boundary detection in digital image analysis. *Int. J. Miner. Process.* **2011**, *101*, 28–36. [CrossRef]
10. Patel, A.K.; Chatterjee, S.; Gorai, A.K. Development of a machine vision system using the support vector machine regression (SVR) algorithm for the online prediction of iron ore grades. *Earth Sci. Inform.* **2019**, *12*, 197–210. [CrossRef]
11. Patel, A.K.; Chatterjee, S.; Gorai, A.K. Development of machine vision-based ore classification model using support vector machine (SVM) algorithm. *Arab. J. Geosci.* **2017**, *10*, 1–16. [CrossRef]
12. Ma, X.-M. A Revised Edge Detection Algorithm Based on Wavelet Transform for Coal Gangue Image. In Proceedings of the 2007 International Conference on Machine Learning and Cybernetics, Hong Kong, China, 19–22 August 2007; pp. 1639–1642.
13. Xu, D.; Chen, X.; Xie, Y.; Yang, C.; Gui, W. Complex networks-based texture extraction and classification method for mineral flotation froth images. *Miner. Eng.* **2015**, *83*, 105–116. [CrossRef]
14. Chatterjee, S.; Bandopadhyay, S.; Machuca, D. Ore grade prediction using a genetic algorithm and clustering based ensemble neural network model. *Math. Geosci.* **2010**, *42*, 309–326. [CrossRef]
15. Andersson, T.; Thurley, M.J.; Carlson, J.E. A machine vision system for estimation of size distributions by weight of limestone particles. *Miner. Eng.* **2012**, *25*, 38–46. [CrossRef]
16. Delbem, I.; Galéry, R.; Brandão, P.; Peres, A. Semi-automated iron ore characterisation based on optical microscope analysis: Quartz/resin classification. *Miner. Eng.* **2015**, *82*, 2–13. [CrossRef]
17. Xiao, D.; Liu, X.; Le, B.T.; Ji, Z.; Sun, X. An ore image segmentation method based on RDU-Net model. *Sensors* **2020**, *20*, 4979. [CrossRef] [PubMed]
18. Liu, Y.; Zhang, Z.; Liu, X.; Wang, L.; Xia, X. Efficient image segmentation based on deep learning for mineral image classification. *Adv. Powder Technol.* **2021**, *32*, 3885–3903. [CrossRef]
19. Baraian, A.; Kellokumpu, V.; Paaso, J.; Koresaar, L.; Kaartinen, J. Computing Particle Size Distribution of Mineral Rocks Using Deep Learning-Based Instance Segmentation. In Proceedings of the 2022 10th European Workshop on Visual Information Processing (EUVIP), Hong Kong, China, 19–22 August 2022; pp. 1–6.
20. He, K.; Gkioxari, G.; Dollár, P.; Girshick, R. Mask r-cnn. In Proceedings of the IEEE International Conference on Computer Vision, Venice, Italy, 22–29 October 2017; pp. 2961–2969.
21. Wu, S.; Wang, Q.; Zeng, Q.; Zhang, Y.; Shao, Y.; Deng, F.; Liu, Y.; Wei, W. Automatic extraction of outcrop cavity based on a multiscale regional convolution neural network. *Comput. Geosci.* **2022**, *160*, 105038. [CrossRef]
22. Hu, J.; Shen, L.; Sun, G. Squeeze-and-Excitation Networks. In Proceedings of the IEEE Conference on Computer Vision and Pattern Recognition, Salt Lake City, UT, USA, 18–22 June 2018; pp. 7132–7141.
23. Hou, Q.; Zhou, D.; Feng, J. Coordinate Attention for Efficient Mobile Network Design. In Proceedings of the IEEE/CVF Conference on Computer Vision and Pattern Recognition, Online, 19–25 June 2021; pp. 13713–13722.

24. Jaderberg, M.; Simonyan, K.; Zisserman, A. Spatial transformer networks. In *Advances in Neural Information Processing Systems Proceedings of the NIPS 2015, Montreal, Canada, 7–12 December 2015*; Neural Information Processing Systems Foundation, Inc (NeurIPS): La Jolla, CA, USA, 2015; pp. 1–9.
25. Yu, F.; Koltun, V.; Funkhouser, T. Dilated Residual Networks. In Proceedings of the IEEE Conference on Computer Vision and Pattern Recognition, Venice, Italy, 22–29 October 2017; pp. 472–480.
26. Li, X.; Sun, X.; Meng, Y.; Liang, J.; Wu, F.; Li, J. Dice loss for data-imbalanced NLP tasks. *arXiv* **2019**, arXiv:1911.02855.
27. Wu, S.; Li, X.; Wang, X. IoU-aware single-stage object detector for accurate localization. *Image Vis. Comput.* **2020**, *97*, 103911 [CrossRef]
28. Lin, T.-Y.; Maire, M.; Belongie, S.; Hays, J.; Perona, P.; Ramanan, D.; Dollár, P.; Zitnick, C.L. Microsoft Coco: Common Objects in Context. In Proceedings of the Computer Vision–ECCV 2014: 13th European Conference, Zurich, Switzerland, 6–12 September 2014; pp. 740–755.

water

MDPI

Article

Water Stable Isotopes in the Central Altai Mountainous Rivers as Indicator of Glacier Meltwater Fraction in Runoff

Dmitrii Bantcev [1,*], Valeriia Rasputina [1], Anaiit Ovsepian [1], Semyon Griga [1], Anna Kozachek [1,2], Kirill Tchikhatchev [2] and Dmitrii Ganyushkin [1]

1 Institute of Earth Sciences, St. Petersburg State University, St. Petersburg 199034, Russia; v.rasputina@spbu.ru (V.R.); annahit2002@mail.ru (A.O.); semyon.griga@yandex.ru (S.G.); kozachek@aari.ru (A.K.); ganushkinspbgu@mail.ru (D.G.)

2 Arctic and Antarctic Research Institute, St. Petersburg 199397, Russia; tchikhatchev@gmail.com

* Correspondence: bancev-d@yandex.ru

Abstract: We used stable water isotopes (δ^{18}O and δ^{2}H) to identify the fractions of glacier meltwater and summer precipitation in the runoff in the Taldura River in the Altai mountains. The mean isotopic characteristics of glacier ice, snow, summer precipitation and river water were obtained. Using isotopic separation of hydrographs, we determined that glacier feeding completely prevails throughout the Taldura River in the middle of the ablation season. In general, the fraction of glacier meltwater in the Taldura River's runoff in the ablation season varies from 80% to 95% depending on local weather conditions.

Keywords: stable isotopes; glacier meltwater; glacier rivers; hydrograph separation

1. Introduction

The main purpose of this study is the estimation of the glacier meltwater fraction in runoff in the Taldura river basin using stable isotope data. This research was carried out on the South-Chuya ridge. The hydrological knowledge of this ridge is extremely insufficient. Currently, most of the gauge stations in the entire Altai Mountains are closed. Closest to the ridge is the Kucherla station on the Kucherla River, located 107 km northwest of the Katunsky ridge in completely different conditions: higher maximal elevation of the ridge, larger amount of precipitation and larger glaciation area. In the Altai region, glacier-originated rivers have a great impact on the economic activities of the local population; therefore, an estimate of the contribution of modern glaciation and other water sources to the runoff is necessary to assess possible changes in the water balance under the conditions of changing climate and deglaciation.

Water stable isotopes (^{18}O and ^{2}H) are currently widely used in upland hydrology. These isotopes in general are used as tracers to identify and estimate water sources. This kind of research is widespread in mainly glaciated mountain territories, especially in the Himalayas [1]. However, such studies have not been carried out practically for the Russian Altai. The complex isotope hydrological studies closest to Altai were carried out on the territory of China at the Urumqi river basin [2], on the Tibetan Plateau [3–6] and in mountainous areas of southwestern China [7]. Isotope studies in Altai are primarily associated with the deep-core drilling of glaciers on the Belukhas plateau [8] and at the Tsambagarav massif in Northwestern Mongolia [9]. For the Altai Mountains and its foothills, isotope analysis of precipitation was performed and compared with model data [10–12].

Earlier, we performed isotope studies at the Tavan-Bogdo-Ola, Tsambagarav and Mongun-Taiga mountain ranges [13,14]. In 2022, the first isotope samples were taken in the Taldura river basin [15].

The Taldura drainage basin belongs to the upper reaches of the unified Ob-Irtysh river system. The isotopic composition of the Ob River in the city of Barnaul was recently investigated [16], but at the same time, isotope data are insufficient in the high mountainous parts of the Ob river basin.

2. Materials and Methods

2.1. Study Area

Isotope hydrological studies were carried out in July–August 2023 in the Taldura River catchment area located in the upper reaches of the Chuya river basin, which is part of the Ob′ river basin. The Taldura river originates from the glaciers on the South-Chuya ridge. This ridge is located in the Central Altai and extends from west to east for 120 km. The highest elevation is 3967 m. The South-Chuya ridge is characterized by a large number of glaciers: the total glaciation area is the second largest in the Russian Altai (Figure 1).

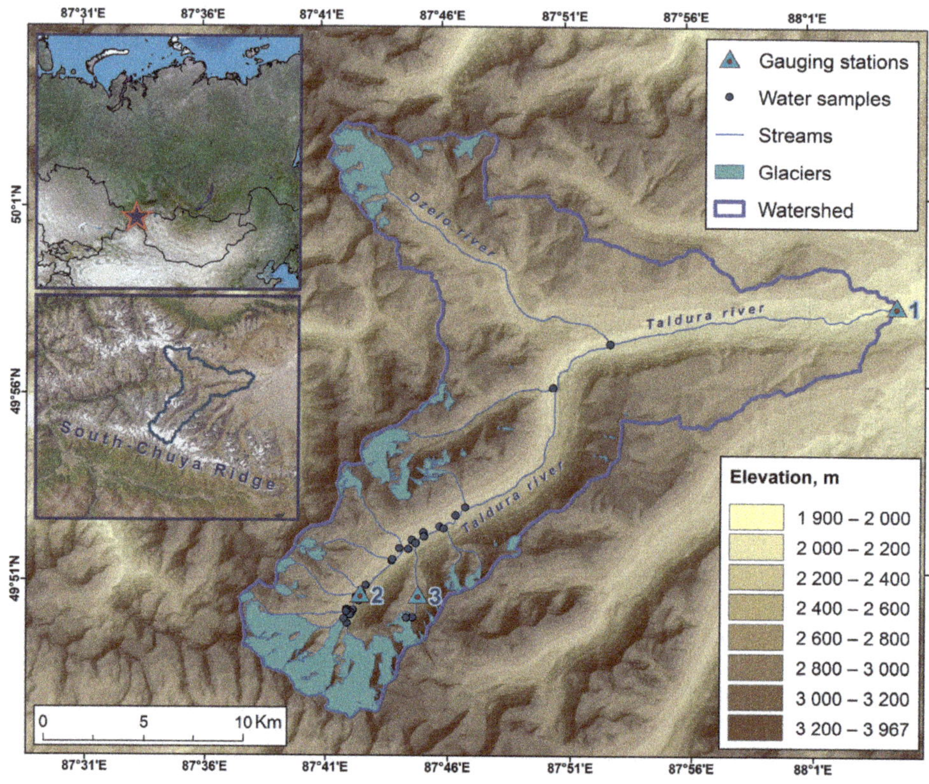

Figure 1. Map of the study area. 1—Lower gauging station. 2—Upper gauging station. 3—Additional gauging station on the tributary.

As a result of climate change, which according to [17] in the territory of the Russian Federation began in the late 1970s, the glaciers of the South-Chuya ridge are shrinking. At the moment, the glaciation area of the South-Chuya Ridge continues to decrease. The total area of glaciers is estimated as 118 km^2. Around 70 glaciers with a total area of 39.8 km^2 are located in the Taldura river basin. The equilibrium line elevation is 3000–3100 m [18]. The Taldura River originates from the Bolshaya Taldura glacier. The river valley is located on the northern macroslope of the South-Chuya ridge (Figure 1). According to the Catalogue of Glaciers of Russia [19], the Bolshaya Taldura glacier system had an area of 20.46 km^2 with a maximum length of 4760 m, a vertical range of 1240 m and an average height of the firn line of 3120 m in 2017.

In this part of Altai, the annual precipitation amount is 400–600 mm. On the windward slopes, it can reach 800–1000 mm. From April to October, 70–85% of the annual precipitation falls. Its monthly maximum is usually observed in June–July [20].

The area of the studied part of the Taldura river basin is 347.8 km^2 (Figure 1). The total glaciation of this part of the basin was estimated at 11.4%.

2.2. Methods

Sample collection was carried out in the middle of the ablation season, from July to August. Most of the samples were taken at the 3 temporary gauging stations (Figure 1). The lower Taldura gauging station was established at an altitude of 1936 m. The upper Taldura gauging station was established at an altitude of 2420 m at a distance of 1600 m from the edge of the Bolshaya Taldura glacier. The observation period at the upper station was 28 days, at the lower one was 18 days. On the right tributary of the Taldura River, at an altitude of 2760 m, the additional gauging station was established. The observation period was 24 days.

At each gauge, the water level was measured 4 times a day and the discharge was also measured periodically. The stage–discharge relation was calculated and for each value of water level, the discharge value was obtained.

At all gauging stations, meteorological observations were also carried out. The air temperature was recorded using the EClerk-M-RHT automatic recorder [21]. The amount of precipitation was determined using a portable rain gauge.

Total dissolved solids (TDS) in water samples was measured by a HANNA HI 98311 Dist5 conductivity meter [22].

Isotope samples at the gauging stations were taken in 50 mL test tubes twice a day: at 08:00 and 20:00. In cases of intense precipitation or high melting, an additional sample was taken, usually at 16:00. Samples of glacier ice, snow, precipitation and different streams were also taken. Glacier ice was collected from the surface of glaciers into plastic bags, then melted at the surrounding temperature and poured into test tubes. The samples from the streams were taken directly into the test tubes. Precipitation was usually sampled twice a day (08:00 and 20:00) from a portable rain gauge.

The relative isotopic composition is indicated by the δ SMOW (standard mean ocean water) and is accepted as the zero standard [23].

$$\delta^{18}O = ([(^{18}O/^{16}O)_{sample} - (^{18}O/^{16}O)SMOW]/(^{18}O/^{16}O)_{SMOW}) \times 10^3 \quad (1)$$

$$\delta^{2}H = ([(^{2}H/H)_{sample} - (^{2}H/H)SMOW]/(^{2}H/H)_{SMOW}) \times 10^3 \quad (2)$$

The analysis of isotopic characteristics was carried out at the Laboratory for Climate and Environmental Change in the Arctic and Antarctic Research Institute (St. Petersburg, Russia) by Cavity Ring-Down Spectroscopy (CRDS) using Picarro L2130-i and Picarro L2140-I analyzers [24]. Distilled tap water from St. Petersburg was used as a laboratory standard, and it had the following characteristics: –9.66 ‰ in $\delta^{18}O$ and –74.1 ‰ in δ^2H relative to the IAEA "V-SMOW2" standard. The measurement precision was 0.04 ‰ for $\delta^{18}O$ and 0.4 ‰ for δ^2H, which is sufficient for this kind of research.

Using stable isotopes as tracers, it is possible to divide the hydrograph into components. The use of stable isotopes in glaciology and hydrology is based on natural differences in the isotopic composition of the water from different sources, e.g., glacier meltwater, groundwater and precipitation. Using only $\delta^{18}O$ values, only two-component separation can be carried out using the isotope balance formula, which generally has the following form:

$$R^{18}O_1f_1 + R^{18}O_2f_2 = R^{18}O \quad (3)$$

with $R^{18}O_1$—isotopic content of the first component; f_1—fraction of the first component; $R^{18}O_2$—isotopic content of the second component; f_2—fraction of the second component; $R^{18}O$—resulting isotopic composition [25].

3. Results

3.1. Isotopic Composition of Precipitation

Precipitation samples were taken at different altitudinal levels: at each gauging station and additionally from the surface of the corrie glacier near gauge no. 3 at an elevation of 3000 m. The isotopic composition of the precipitation at every sampling point is shown in Figure 2.

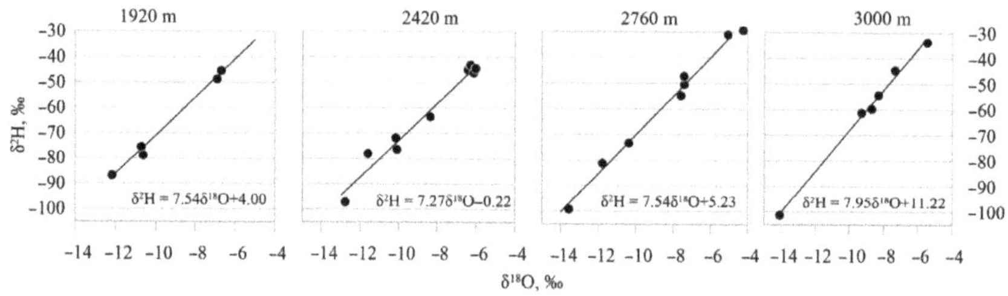

Figure 2. Isotopic composition of the summer precipitation of 2023 at different altitudinal levels.

The local meteoric water line (LMWL) was plotted from the event-based precipitation samples for each elevation separately. The slope of these lines is lower than that of the GMWL, which is common for Inner Asia summer precipitation [13,14,26,27]. The LMWL of the samples from the glacier surface (Figure 2, 3000 m) is close to the GMWL. The weighted-average $\delta^{18}O$ of the samples ranges from −8.3‰ to −10.4‰.

The weighted-average $\delta^{18}O$ values of ^{18}O were compared with the modeled values from the online calculator OIPC (The Online Isotopes in Precipitation Calculator), which allows the monthly or annual isotopic composition to be calculated using geographical coordinates and elevation (Table 1). This calculator is based on the interpolation of IsoMAP data based on GNIP (Global Network of Isotopes in Precipitation) weather stations [28–30].

Table 1. Comparison of weighted-average $\delta^{18}O$ of precipitation in July–August 2023 and OIPC model data (Bowen et al., 2005, Bowen, 2017) [28,29].

	Weighted-Average $\delta^{18}O$, ‰	$\delta^{18}O$ in July by OIPC, ‰
Gauge 1, 1920 m	−10.4	−8.3
Gauge 2, 2420 m	−8.3	−8.8
Gauge 3, 2760 m	−9.8	−9.2
Glacier near the gauge 3, 3000 m	−9.3	−9.6

The modeled monthly $\delta^{18}O$ values are rather close to obtained $\delta^{18}O$, except at the lowest sampling point. The isotopic composition of precipitation (Figure 2, Table 1) also shows a lack of altitude effect (decrease in $\delta^{18}O$ with increasing altitude) in July 2023. Since the altitude effect was absent or weakly expressed in the precipitation, differences in the isotopic composition of precipitation at different altitudinal levels can be ignored in the process of hydrograph separations during summer 2023.

3.2. Isotopic Composition of Ice and Snow

Ice samples were taken from the ablation zone of the surface of the Bolshaya Taldura glacier at relatively regular intervals. This technique has been used before and has proven itself fit for determining the average isotopic composition of ice [31]. (Seasonal snow was collected from two snow pits in snow fields on the surface of the glacier tongue. In total, 45 ice and 17 snow samples were collected.) The sampling points and the isotopic composition of the ice and snow are shown in Figure 3.

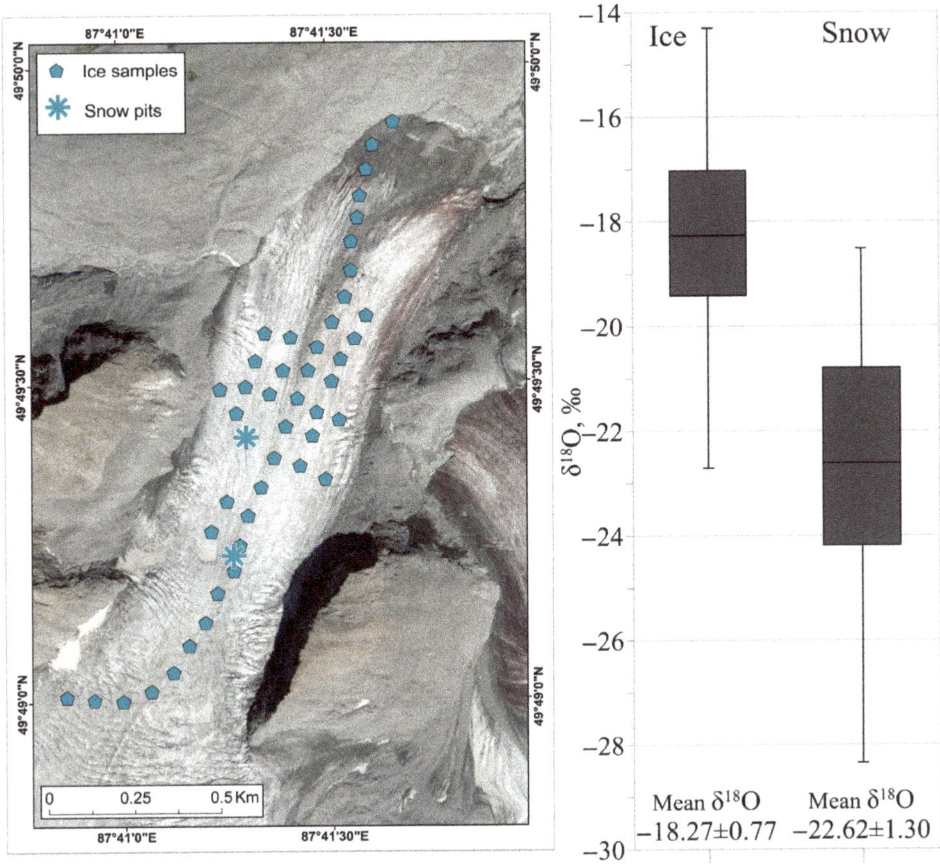

Figure 3. Sampling points on the surface of the Bolshaya Taldura glacier and $\delta^{18}O$ box plots for the glacier ice and snow.

The mean and range of $\delta^{18}O$ in glacier ice indicate the formation of ice from precipitation of the transitional seasons (autumn–spring). In the seasonal snow on the glacier surface, lower values of $\delta^{18}O$ are presented, so we can indicate this snow as winter snow from 2022–2023. The meteoric lines in the ice and snow are close to each other and close to the GMWL: $\delta^2H = 7.82 \times \delta^{18}O + 13.29$ in ice and $\delta^2H = 7.78 \times \delta^{18}O + 9.38$ in snow. These meteoric lines are almost very close to the meteoric line ($\delta^2H = 7.8 \times \delta^{18}O + 7.7$) obtained from 524 samples of firn core on the Belukha plateau in 2002 [8]. A summary of the ice and snow characteristics is shown in Table 2.

Table 2. Characteristics of the ice and snow from Bolshaya Taldura glacier.

Type	Number of Samples	Range of $\delta^{18}O$, ‰	Mean $\delta^{18}O$, ‰	TDS, ppm	Meteoric Water Line
Ice	45	−26.67–−12.31	−18.27 ± 0.77	2.5 ± 0.8	$\delta^2H = 7.82 \times \delta^{18}O + 13.29$
Snow	17	−18.52–−28.34	−22.62 ± 1.30	8.4 ± 5.4	$\delta^2H = 7.78 \times \delta^{18}O + 9.38$

The ice is characterized by very low values of total dissolved solids (TDS). In the snow, the mean TDS is 8.4 ppm; this value is also quite low. The high confidence interval is explained by the smaller number of snow samples taken.

3.3. Isotopic Composition of Water

For every observation day at each gauging station, the mean $\delta^{18}O$ was calculated. Figure 4 shows that the variability in $\delta^{18}O$ at the lower hydrological post is less than that at the upper one, while the courses of the $\delta^{18}O$ values at the two gauges for the total period coincide. Precipitation at the lower post practically has no effect on the isotopic composition of the water.

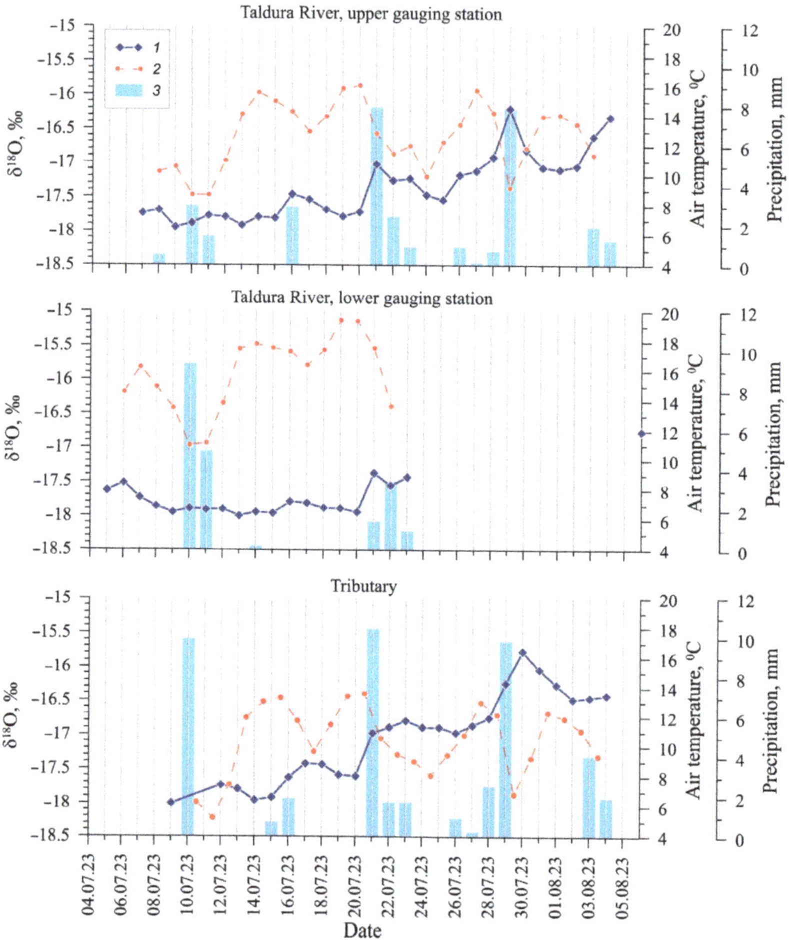

Figure 4. Changes in $\delta^{18}O$ and meteorological parameters at gauging stations. 1—$\delta^{18}O$; 2—temperature; 3—precipitation.

The increase in $\delta^{18}O$ at each gauge is associated with precipitation. At the same time, the highest value of $\delta^{18}O$ at the upper gauging station was recorded on 29.07, when a large amount of precipitation occurred at the period of low air temperature and, consequently, minimal melting. In general, there was a tendency for an increase in $\delta^{18}O$ at the upper gauging station after 21.06.

At the lower gauging station, the situation is uncommon: the $\delta^{18}O$ values in general are lower than at the upper one, which indicates a greater fraction of glacier meltwater to the runoff, despite the distance of 34 km from the edge of the Bolshaya Taldura glacier. But a separate sample taken at the lower post on 06.08 showed that there was a tendency for

$\delta^{18}O$ to increase, as well as at the upper post. There is also minimal variability in $\delta^{18}O$ at the lower gauge during the observation period.

At the gauging station on the tributary, the range of $\delta^{18}O$ is greater and the mean $\delta^{18}O$ is higher, which indicates a greater influence of precipitation on the runoff in the small tributary. As the other two gauges, there is trend of an increase in $\delta^{18}O$ over time and the peaks of $\delta^{18}O$ after precipitation events are pronounced.

Samples were also taken from different tributaries of both glacial and non-glacial origin. The water characteristics obtained from the gauges and tributaries are shown in Table 3.

Table 3. Average characteristics of water samples by group.

Type of the Samples	Mean Discharge, m^3/s	Mean $\delta^{18}O$, ‰	Mean TDS, ppm
Gauge 1, lower Taldura	25.0	−17.77	41.9
Gauge 2, upper Taldura	10.95	−17.30	3.3
Gauge 3, tributary	1.08	−17.05	7.8
Glacier-originated tributaries	-	−17.58	18.4
Non-glacier-originated tributaries	-	−15.77	36.3

The obtained mean value and the range of $\delta^{18}O$ (from −18.79 to −16.57‰) of water from glacier originated tributaries are very close to the isotopic composition of water at the gauging stations on the Taldura River. It can be suggested that these values are typical for the glacier rivers of the ridge in the middle of the ablation season.

The mean $\delta^{18}O$ value in the streams of non-glacial origin is −15.77‰ (from −17.11 to −14.55‰). The relatively low isotopic composition indicates the contribution of snowmelt water to their feeding.

In the upper reach of the Taldura River (gauge 2), the TDS changes slightly during the observation period (within 2–4 ppm). At the lower reach of the Taldura river (gauge 1), the TDS was 10 times higher because of enrichment from the dissolved mineral compounds. On the right tributary of the Taldura River, the TDS remains almost unchanged throughout the observation period (7–8 ppm).

The TDS in the glacier-originated tributaries is lower than the mineralization of non-glacial originated tributaries but higher than the water conductivity at gauges 2 and 3, because of the filtration through moraine.

3.4. Isotopic Separation

The isotopic composition of the water on the gauges of the Taldura River reflects a mixture of water from different sources (Figure 5). The positions of the gauges on this diagram show the dominance of meltwater in river feeding.

According to the isotope balance equation, Equation (3), a two-component separation of hydrographs was performed. Several assumptions have been made to facilitate the separation. Only glacier meltwater and precipitation were considered components because it is possible to separate only two components using only $\delta^{18}O$ as a tracer. It was also assumed that there is no separate groundwater reservoir in the studied basin, which differs in isotopic composition from summer precipitation and meltwater. This fact is confirmed by the absence of significant springs forming large tributaries in the Taldura river basin.

The average value $\delta^{18}O$ of the glacial ice of the Bolshaya Taldura glacier was taken as the isotopic composition of meltwater. Precipitation was considered the second component, since the trends in the isotopic composition of water at all gauging stations showed that precipitation affects changes in the isotopic composition of river water. The isotopic composition of precipitation was taken as the weighted-average value of $\delta^{18}O$ for the day before sampling, and the daily course of the isotopic composition of water showed that the run-up time did not exceed 24 h. If no precipitation occurred, the separation was performed using the isotopic composition of the last precipitation event. The results of isotopic separation at the Taldura River gauging stations are shown in Figure 6.

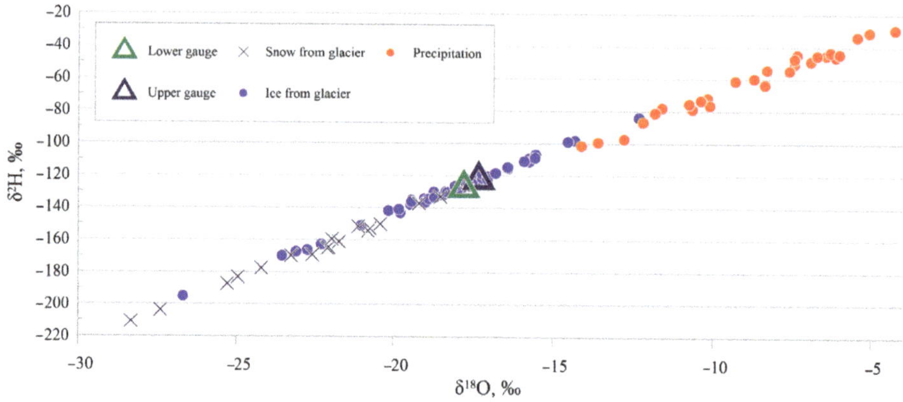

Figure 5. $\delta^{18}O$–$\delta^{2}H$ diagram for different groups of the samples.

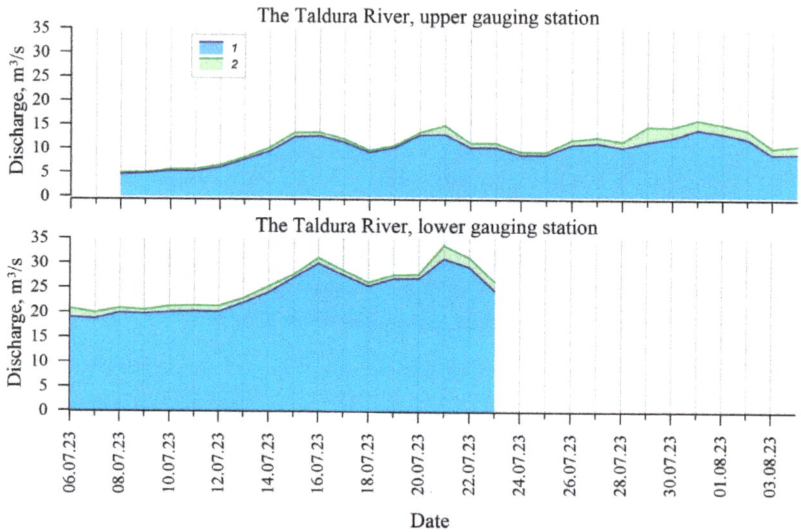

Figure 6. Separated hydrographs of the Taldura River: 1—meltwater of glaciers; 2—precipitation.

At the upper hydrological gauging station, the fraction of precipitation in runoff varied from 4 to 22% and averaged 9%. The average daily discharge formed from atmospheric precipitation varied from 0.18 to 3.19 m^3/s. The maximum fraction of precipitation was observed on 29–30 July after a large amount of precipitation, with a decrease in temperature and, as a result of the low temperature, a decrease in melting.

At the lower gauging station, over a shorter observation period (until 23.07), the fraction of precipitation varied from 2 to 9%, with an average precipitation fraction of 5%. The average daily discharge formed from precipitation varied from 0.64 to 2.70 m^3/s. During this period the discharges formed by precipitation in the upper gauging station were 0.18–1.88 m^3/s. This fact may indicate that the isotopic separation results are correct. The maximum precipitation fraction in the runoff was observed on 21–23 July after precipitation in the upper part of the basin. It was also found that on 06.08 (based on an additional sample, not included in Figure 5), the fraction of precipitation increased to 17%.

The tributary (gauge 3) is characterized by an order of magnitude lower discharge. It was found that the fraction of precipitation varied from 5 to 32% (on average 15%), which is greater than on the Taldura River, but it is still mostly insignificant.

4. Discussion

First, it is necessary to sum up the obtained data on the isotopic composition of precipitation. The altitude effect of changes in the isotopic composition of precipitation is not pronounced: a decrease in the values of $\delta^{18}O$ with an increase in altitude is not observed. The absence of an altitude effect can be explained by precipitation from a single condensation level when the air mass does not rise up along the valley or slopes. The reverse altitude effect (minimal $\delta^{18}O$ at the lower gauge) may be associated with various sources of moisture and the turbulent mixing of air masses on orographic barriers [32].

A comparison of precipitation $\delta^{18}O$ with OIPC data showed a close agreement. Therefore, the data from this online calculator may be used for further large-scale isotope hydrological studies.

The isotopic composition of the ice from Bolshaya Taldura can be used in further isotopic separation research in the upper Chuya basin as the isotopic composition of meltwater. The obtained mean $\delta^{18}O$ value is close to the isotopic composition of ice in the Eastern Altai [14]. In the Western Altai, glacier ice must be enriched with stable isotopes according to the greater fraction of summer snow in glacier feeding. This is confirmed by ice drilling results on the Belukha plateau [8].

Changes in the water $\delta^{18}O$ at each gauge reflect the influence of precipitation and changes in melting intensity. The low variability in $\delta^{18}O$ in the lower gauge is explained by averaging the isotopic characteristics of all the water in the basins studied. Small tributaries are more affected by precipitation and temperature changes. This is reflected in the greater variability in water $\delta^{18}O$.

It is interesting that the fraction of glacier feeding in the Taldura River increases with distance from the Bolshaya Taldura glacier. The decrease in the precipitation fraction in the Taldura River from the upper part of the basin to the lower part is explained by the influence of glacier-originated tributaries with a natural decrease in precipitation in the lower part of the basin. The waters of the tributaries, as mentioned above, have an average value of $\delta^{18}O$ −17.61‰. Some of them are characterized by the dominance of glacier meltwater with low $\delta^{18}O$ values in water, e.g., the mean $\delta^{18}O$ from three samples in the Dzelo River (see Figure 1) is −18.17 ‰. In addition, the amount of precipitation in the lower part of the catchment area is naturally decreasing. Thus, glacial feeding prevails completely in the runoff in the Taldura River throughout its entire length and in the runoff in its main tributaries.

Judging by the low values of $\delta^{18}O$ at the start of the observation period at all gauging stations (Figure 4), we can suggest that snow meltwater, which is characterized by the lowest $\delta^{18}O$ (Table 2), contributed to the feeding of rivers. The hydrograph can be divided into three components (glacier meltwater, snow meltwater and precipitation) using two tracers. TDS is usually used as the second one, but in this case the values of the TDS of the ice, snow and precipitation were too close to each other. It is also possible to perform three-component separation using excess deuterium [1]. But for a correct three-component separation, we need more data on the isotopic composition of snow from the glaciers and snowfields over the entire catchment area. In any case, the low values of $\delta^{18}O$ in the lower reaches of the Taldura River can also be explained by an increase in the fraction of snow meltwater.

It is also interesting to investigate the interannual variability in the isotopic composition. Some samples were taken from the Taldura River a year before this research, in 2022. In 2022 and 2023, sampling was carried out in approximately one part of the ablation season (mainly in July), while various meteorological conditions were observed, which makes it possible to assess the variability in the contribution of glacier runoff to the feeding of the

Taldura River. A comparison of isotopic characteristics and meteorological conditions for the example of the upper gauging station is shown in Table 4.

Table 4. Comparison of data from 2022 and 2023 for the upper gauging station at the Taldura River.

Year	2022	2023
Observation period	11.07–29.07	11.07–29.07
Range of $\delta^{18}O$ in Taldura River, ‰	−15.05–−16.92	−16.06–−18.04‰
Mean $\delta^{18}O$, ‰ (upper gauge)	−16.22	−17.30
Mean T, °C (upper gauge)	+9.1	+12.7
Precipitation, mm	38	33

In 2022, the colder conditions with more precipitation, as a result of which the fraction of glacier runoff in the Taldura River's feeding was lower, were reflected in the isotopic composition. At the same time, for the precipitation in the Taldura River, the fraction in 2022 generally did not exceed 20%. Thus, even under different conditions in the middle of the ablation season, glacier meltwater completely prevails in Taldura River feeding.

5. Conclusions

Based on the results of a complex of field work and laboratory measurements, it was determined that glacier feeding completely prevails throughout the Taldura River in the middle of the ablation season. At the same time, there is even a slight increase in the fraction of glacier meltwater in the lower part of the basin, far from the Bolshaya Taldura glacier. The meltwater fraction at the mouth of the Taldura River can reach 95%. Near the source of the Taldura River, the meltwater fraction in general is around 80–95%. The second most important source of water is summer precipitation, but its contribution, depending on the conditions, usually does not exceed 10–20%. Thus, further degradation of the South-Chuya ridge's glaciation will inevitably affect the runoff in the Taldura River and the water balance of the upper Chuya basin as a whole. Now, the role of glacier runoff in the Taldura River is very high, but with further retreat of the glaciers, in addition to a reduction in their area, the intensity of their melting will decrease, since the edges of the glaciers will be higher. Consequently, in the near future, glacier runoff may even increase, but with continued degradation of glaciation, it will inevitably decrease, which will lead to a significant decrease in runoff in the Taldura River.

Author Contributions: Methodology, D.B. and A.K.; formal analysis, A.O.; investigation, D.B., S.G., A.K. and K.T.; writing—original draft preparation, D.B.; writing—review and editing, V.R. and A.K.; visualization, V.R. and A.O.; project administration, D.G.; funding acquisition, D.B. All authors have read and agreed to the published version of the manuscript.

Funding: This research was funded by the Russian Science Foundation, grant number 23-27-00173. «Estimation of various components contribution to the glacial streams runoff in the Southeastern Altai according to isotope indicators».

Data Availability Statement: Isotopic data used for hydrographs separation is available at https://disk.yandex.ru/i/VvaJvO-nlLRbuw.

Conflicts of Interest: The authors declare no conflict of interest.

References

1. Jeelani, G.; Lone, S.A.; Nisa, A.U.; Deshpande, R.D.; Padhya, V. Use of stable water isotopes to identify and estimate the sources of groundwater recharge in an alluvial aquifer of Upper Jhelum Basin (UJB), Western Himalayas. *Hydrol. Sci. J.* **2021**, *66*, 2330–2339. [CrossRef]
2. Sun, C.; Li, W.; Chen, Y.; Li, X.; Yang, Y. Isotopic and hydrochemical composition of runoff in the Urumqi River, Tianshan Mountains, China. *Environ. Earth Sci.* **2015**, *74*, 1521–1537. [CrossRef]
3. Fan, Y.; Chen, Y.; He, Q.; Li, W.; Wang, Y. Isotopic Characterization of River Waters and Water Source Identification in an Inland River, Central Asia. *Water* **2016**, *8*, 286. [CrossRef]

4. Li, Z.; Feng, Q.; Liu, W.; Wang, T.; Guo, X.; Li, Z.; Gao, Y.; Pan, Y.; Guo, R.; Jia, B.; et al. The stable isotope evolution in Shiyi glacier system during the ablation period in the north of Tibetan Plateau, China. *Quat. Int.* **2015**, *380–381*, 262–271. [CrossRef]
5. Wang, C.; Dong, Z.; Qin, X.; Zhang, J.; Du, W.; Wu, J. Glacier meltwater runoff process analysis using δD and δ18O isotope and chemistry at the remote Laohugou glacier basin in western Qilian Mountains, China. *J. Geogr. Sci.* **2016**, *26*, 722–734. [CrossRef]
6. Zhao, L.J.; Yin, L.; Xiao, H.L.; Cheng, G.; Zhou, M.; Yang, Y.; Li, C.; Zhou, J. Isotopic evidence for the moisture origin and composition of surface runoff in the headwaters of the Heihe River basin. *Chin. Sci. Bull.* **2011**, *56*, 406–415. [CrossRef]
7. Meng, Y.; Liu, G.; Li, M. Tracing the Sources and Processes of Groundwater in an Alpine Glacierized Region in Southwest China: Evidence from Environmental Isotopes. *Water* **2015**, *7*, 2673–2690. [CrossRef]
8. Aizen, V.; Aizen, E.; Fujita, K.; Nikitin, S.; Kreutz, K.; Takeuchi, L. Stable-isotope time series and precipitation origin from firn-core and snow samples, Altai glaciers, Siberia. *J. Glaciol.* **2005**, *51*, 637–654. [CrossRef]
9. Herren, P.A.; Eichler, A.; Machguth, H.; Papina, T.; Tobler, L.; Zapf, A.; Schwikowski, M. The onset of Neoglaciation 6000 years ago in western Mongolia revealed by an ice core from the Tsambagarav mountain range. *Quat. Sci. Rev.* **2013**, *69*, 59–68. [CrossRef]
10. Malygina, N.S.; Eirikh, A.N.; Kurepina, N.Y.; Papina, T.S. Isotopic composition of precipitation in Altai foothills: Observation and interpolation data. *Bull. Tomsk. Polytech. Univ.* **2019**, *330*, 44–54. (In Russian) [CrossRef]
11. Malygina, N.S.; Eirikh, A.N.; Kurepina, N.Y.; Papina, T.S. Isotope composition of winter precipitation and snow cover in the foothills of the Altai. *Ice Snow* **2017**, *57*, 57–68. (In Russian) [CrossRef]
12. Papina, T.S.; Malygina, N.S.; Blyakharchuk, T.A.; Nenasheva, G.I.; Ryabchinskaya, N.A.; Eyrikh, A.N. Isotopic composition and pollen spectra of precipitation and ice at the edge part of Korumdu Glacier (North Chu ridge, Altai Mountains). *Ice Snow* **2015**, *55*, 40–48. (In Russian) [CrossRef]
13. Bantcev, D.V.; Ganyushkin, D.A.; Chistyakov, K.V.; Volkov, I.V.; Ekaykin, A.A.; Veres, A.N.; Tokarev, I.V.; Shtykova, N.B.; Andreeva, T.A. The Components of the Glacial Runoff of the Tsambagarav Massif from Stable Water Isotope Data. *Geosciences* **2019**, *9*, 297. [CrossRef]
14. Bantsev, D.V.; Ganyushkin, D.A.; Chistyakov, K.V.; Ekaykin, A.A.; Tokarev, I.V.; Volkov, I.V. Formation of glacier runoff on the northern slope of Tavan Bogd mountain massif based on stable isotopes data. *Ice Snow* **2018**, *58*, 333–342. (In Russian) [CrossRef]
15. Bantcev, D.V.; Ovsepyan, A.A.; Rasputina, V.A.; Kozachek, A.V.; Tchikhatchev, K.B.; Ganyushkin, D.A. The Use of Isotope Indicators in Hydrological Studies in the Taldura River Basin, Central Altai. *Ice Snow* **2023**, *63*, 513–524. (In Russian) [CrossRef]
16. Papina, T.; Eirikh, A.; Kotovshchikov, A.; Noskova, T. Impact of Snowmelt Conditions on the Isotopic Composition of the Surface Waters of the Upper Ob River during the Flood Period. *Water* **2023**, *15*, 2096. [CrossRef]
17. Akentyeva, E.M.; Anisimov, O.A.; Bardin, M.Y.; Zhuravlev, S.A.; Kattsov, V.M.; Kiselev, A.A.; Klyueva, M.V.; Konstantinov, P.I.; Korotkov, V.N.; Kostyanoy, A.G.; et al. *The Third Assessment Report of Roshydromet on Climate Change and Its Consequences on the Territory of the Russian Federation: General Summary*; Naukoemkie Tehnologii: Saint-Petersburg, Russia, 2022. (In Russian)
18. Okishev, P.A. *Relief and Glaciation of Russian Altai*; Tomsk University Publishing House: Tomsk, Russia, 2011. (In Russian)
19. Khromova, T.Y.; Nosenko, G.A.; Glazovsky, A.F.; Muraviev, A.Y.; Nikitin, S.A.; Lavrentiev, I.I. New Inventory of the Russian glaciers based on satellite data (2016–2019). *Ice Snow* **2021**, *61*, 341–358. (In Russian) [CrossRef]
20. *Resources of Surface Waters of the USSR*; Responsible editor Semenov V.A.; Hydrometeoizdat: Leningrad, Russia, 1969; Volume 15. (In Russian)
21. Temperature and Humidity Logger with Indication EClerk-M-RHT. Available online: https://relsib.com/product/logger-vlazhnosti-i-temperatury-registrator-eclerk-m-rht?ysclid=lvao1tcns292535277 (accessed on 22 April 2024).
22. Hanna Instruments Laboratory and Industrial Analytical Equipment. Available online: https://www.hannarus.ru/ (accessed on 22 April 2024).
23. Reference Materials-VSMOW2. Available online: https://analytical-reference-materials.iaea.org/vsmow2 (accessed on 22 April 2024).
24. Picarro. Available online: https://www.picarro.com/ (accessed on 22 April 2024).
25. Chizhova, Y.u.N.; Rec, E.P.; Vasil'chuk, Y.u.K.; Tokarev, I.V.; Budanceva, N.A.; Kireeva, M.B. Two approaches to hydrograph separation of the glacial river runoff using isotopic methods. *Ice Snow* **2016**, *56*, 161–168. (In Russian) [CrossRef]
26. Saidaliyeva, Z.; Shahgedanova, M.; Yapiyev, V.; Wade, A.; Akbarov, F.; Esenaman, M.; Kapitsa, V.; Kassatkin, N.; Kayumova, D.; Rakhimov, I.; et al. Isotopic composition as a tracer of different source contributions to stream flow in the glacierized catchments of Central Asia. In Proceedings of the Abstracts EGU General Assembly. EGU General Assembly, Vienna, Austria, 24–28 April 2023; EGU23-10181. [CrossRef]
27. Yamanaka, T.; Tsujimura, M.; Oyunbaatar, D.; Gombo, D. Isotopic variation of precipitation over eastern Mongolia and its implication for the atmospheric water cycle. *J. Hydrol.* **2007**, *333*, 21–34. [CrossRef]
28. Bowen, G.J. The Online Isotopes in Precipitation Calculator, Version 3.1. [Online]. 2017. Available online: https://www.waterisotopes.org (accessed on 3 February 2024).
29. Bowen, G.J.; Wassenaar, L.I.; Hobson, K.A. Global application of stable hydrogen and oxygen isotopes to wildlife forensics. *Oecologia* **2005**, *143*, 337–348. [CrossRef]
30. IAEA/WMO. Global Network of Isotopes in Precipitation. The GNIP Database [Online]. 2015. Available online: https://nucleus.iaea.org/wiser (accessed on 3 February 2024).

31. Bantcev, D.; Ganyushkin, D.; Terekhov, A.; Ekaykin, A.; Tokarev, I.; Chistyakov, K. Isotopic Composition of Glacier Ice and Meltwater in the Arid Parts of the Altai Mountains (Central Asia). *Water* **2022**, *14*, 252. [CrossRef]
32. Vasil'chuk, Y.K.; Chizhova, J.N. Altitudinal gradient of δ18O and δD in precipitation and snow cover in high mountains. *Earth's Cryosphere* **2010**, *14*, 13–21.

minerals

MDPI

Article

Extraction Kinetics of Rare Earth Elements from Ion-Adsorbed Underclays

Priscilla Prem [1,2,3,*], Ward Burgess [3,4], Jon Yang [5,6] and Circe Verba [6]

1 Oak Ridge Institute for Science and Education, 100 ORAU Way, Oak Ridge, TN 37830, USA
2 Department of Chemical and Petroleum Engineering, University of Pittsburgh, 4200 Fifth Ave, Pittsburgh, PA 15260, USA
3 National Energy Technology Laboratory, US Department of Energy, 626 Cochran Mill Rd., Pittsburgh, PA 15236, USA; ward.burgess@netl.doe.gov
4 NETL Support Contractor, 626 Cochran Mill Rd., Pittsburgh, PA 15236, USA
5 NETL Support Contractor, 1450 Queen Ave SW, Albany, OR 97321, USA; jonathan.yang@netl.doe.gov
6 National Energy Technology Laboratory, US Department of Energy, 1450 Queen Ave SW, Albany, OR 97321, USA; circe.verba@netl.doe.gov
* Correspondence: priscilla.prem@pitt.edu

Abstract: Citric acid has been identified as an environmentally sustainable organic acid capable of leaching up to ~30% of easily accessible REEs from underclay material. An analysis of the leaching profiles was performed to discern the reaction rates, extraction efficiencies, and potential leaching mechanisms of REEs and cations of interest from ion-adsorbed underclays. The initial leaching stage follows a slow intraparticle diffusion mechanism followed by a second stage controlled by a mixed diffusion regime. The leaching profiles of Ca and P were similar to those of REEs, suggesting that REEs are most likely derived from mineral surfaces such as hydroxyapatite or crandallite rather than predominately from underclays. Fitting to a modified diffusion control model found diffusion-controlled leaching to be the primary mechanism whereas non-diffusive mechanisms made up about 22% of the extracted REEs. Gangue cations associated with underclays had less non-diffusive leaching than REE species, indicating that their leaching kinetics may be dominated by diffusion from within the material or potentially from product layer formation. Fitting to Boyd plots further indicated that REEs were leached following intraparticle diffusion control. These results have important implications for the development of more efficient and sustainable methods for extracting REEs or critical minerals from alternative feedstocks.

Keywords: rare earth elements; citric acid; leaching kinetics; ion-adsorbed underclay; modified diffusion control model

1. Introduction

Rare earth elements (REEs, including the 14 naturally occurring lanthanide elements plus yttrium and scandium) are used in a variety of advanced technology applications such as magnets, catalysts, and other electronic devices. REEs are especially useful for the fabrication of renewable energy technologies and are therefore essential for advancement towards carbon neutrality and sustainable energy [1]. However, the U.S. does not produce enough domestic REEs to support renewable energy generation, let alone enough for all other domestic applications [2]. To satisfy demand, the United States imports approximately 80% of its rare earth compounds and metals from China [3], but the growing need for REE resources highlights the need to find and utilize new domestic sources.

The U.S. controls just 1% of primary REE reserves globally. In fact, REEs extracted from REE-bearing clay deposits are estimated to account for ~35% of China's total REE production despite making up just a small fraction of REE reserves and containing just 0.05–0.2 wt.% rare earth oxides (REOs) [4,5]. This is because the lower amounts of available

REEs are offset by their ease of extraction and processing from ion-adsorbed clays (IACs) [6]. IACs are a potential alternative source of REEs and can be found in mineral deposits adjacent to coal seams and within coal utilization byproducts. The negatively charged surfaces of clays attract positively charged REE cations with the ratio of complexation depending on the pH [7]. Due to their weak ionic interactions, these surface-bound REEs are much easier to procure than the REOs embedded within carbonate and phosphate minerals. Other sources of REEs are carbonatite (source of light REEs), bastnasite deposits, and phosphate-bearing minerals like monazite and xenotime that contain the largest amounts of REOs [5].

Extraction of REEs from carbonate and phosphate minerals often requires the use of aggressive, corrosive mineral acids to dissolve the ore, but these methods require waste management and have numerous environmental concerns for usage [6]. On the other hand, the primary method of REE extraction from IACs utilizes electrolyte solutions such as ammonium sulfate to leach ion-exchangeable REEs bound to the clay surface with 80%–90% extraction efficiencies at moderate temperatures [8]. Chemically, the monovalent ammonium cations exchange with REE cations bound to the surface of IACs in order to release them into the leachate solution. However, even this method risks introducing toxic ammonium and sulfate ions into ground waters during in situ leaching processes [9]. A more environmentally friendly process is to use low molecular weight organic acids capable of complexing directly with REE cations while avoiding both ammonium and sulfate. Citric acid has been previously identified as an effective organic acid to suppress REE adsorption to soils by complexing to the REEs in solution [10]. The chelating properties of the citrate are the result of negatively charged oxygen atoms of the carboxylic groups which complex to REE cations. This increases the solubility of the metals, thereby enabling extraction while offering environmental benefits over mineral acids [11,12]. This process works similarly to glycine lixiviants which have been shown to selectively recover copper from copper smelting slag under basic pH conditions [13].

Yang et al. (2020) characterized various Appalachian underclays and found that the REE-bearing phases consisted of apatite, monazite, xenotime, rhabdophane, and churchite embedded within the clay matrix that were clastic or detrital in origin with alteration products such as crandallite limited to a few feedstocks [14]. Benchtop leaching experiments performed by Montross et al. (2020) successfully leached up to ~30% of exchangeable total REEs (TREEs) from unconventional underclay sediments using a 0.1 M citric acid solution [15]. The current study models the kinetics of data acquired from upscaled experiments performed by Yang et al. (2022) for which the operating conditions were obtained in previous benchtop studies to determine the highest REE extraction efficiencies using citrate-based solutions [16]. However, the REE extraction efficiencies of the upscaled experiments were significantly lower than those of the benchtop leaching experiments performed by Montross et al. (2020) and inspired the current study to determine the kinetics and potential leaching mechanisms in the upscaled experiments. The probable cause for this discrepancy in extraction efficiencies is likely due to the different feedstock compositions and experimental parameters that differ between the benchtop and large-scale experiments. One such concern is the impact of capillary forces which are harder to overcome in upscaled experiments due to restrictions in mixing at large scale. Recent studies aimed at overcoming low extraction efficiencies in acid leaching experiments showed that coupling chemical leaching with metallurgical techniques such as disintegration liberates REE-bearing minerals in clay materials and improves their contact with the solution [17]. Factors influencing the extraction efficiencies between benchtop and upscaled experiments may be better revealed with a deeper understanding of the leaching kinetics and mechanisms.

To better understand these lower extraction efficiencies in the upscaled embodiments, the kinetic profiles can provide insight into the rate determining mechanisms that control the REE leaching process. Examination of the rate determining mechanisms throughout the leaching process is useful for optimizing REE leachability. Once the leaching mechanisms of REEs and relevant cations are determined, correlations to specific phases can be postulated and experimental parameters can be adjusted to selectively leach REE-rich phases over those primarily consisting of unwanted gangue elements such as Al, Si, and Fe. Therefore, REE-rich phases can be selectively targeted for extraction if they exhibit leaching behaviors that can be distinguished from other phases primarily containing other critical minerals (CMs) or gangue elements. This study aims to fit REE and other major cation (e.g., Al, Ca, Fe, P, Si, Co, Ni, Cu) concentration data to commonly used kinetic models to determine whether the rate limiting mechanism is a diffusion or chemically controlled process. Depending on the mechanism, the resulting kinetic parameters can be defined to describe the leaching behavior of different phases.

2. Materials and Methods

2.1. Sample Preparation and Analysis

The experimental data used in this study were derived from Yang et al. (2022) [16]. In summary, bulk underclay samples obtained from the Middle Kittanning coal formation (MKT) in West Virginia were crushed and sieved to produce underclay fractions with an average particle size (diameter) of 1.858 mm (r = 0.0929 cm). To simulate batch leaching, 8.0 kg MKT underclay material was placed in a 55 gal mixing barrel equipped with an axial flow impeller to ensure complete vertical mixing. The mixing barrel was filled with 0.1 M citric acid lixiviant containing both citric acid ($HOC(COOH)(CH_2COOH)_2$) and sodium citrate dihydrate ($HOC(COONa)(CH_2COONa)_2 \bullet 2H_2O$) for a final water-to-rock ratio of 10:1 at pH 2.8 over the course of 18 days at ambient conditions. At the conclusion of the experiment, the solids were rinsed with tap water, collected, and dried. The REE and major cation concentrations in the leachate were collected at several time points and those datasets were used for the modeling described in the next sections; detailed bulk geochemistry is noted in Bauer et al. (2022) and Yang et al. (2022) [16,18].

2.2. Kinetic Modeling Methods

Due to the complexity of the heterogenous underclay material, we took multiple approaches to determine the rate limiting mechanism of the leaching process. The rate controlling mechanism was determined by first fitting extraction data to non-linear kinetic equations for geometric contraction models commonly used to describe the extraction of REEs from mineral sources [19–21]. Table 1 describes several models including the shrinking core model (SCM), Jander equation (JE), and Ginstling–Brounshtein model (GBM). In each of the geometric contraction models, X represents the mass fraction of the species leached from the IAC sample as determined by the initial mass of the species within the ore (M_i) and the mass leached at time t (M_t). M_t was determined by measuring the concentration of REEs and other cations extracted into the leachate, multiplying by the liquid volume, and dividing by the original mass of solid loaded into the reactor. Regression analysis and data fitting to the selected models were performed using GraphPad Prism 9 software and the goodness of fit was evaluated using the R-squared values.

The leaching of REEs from IACs occurs via the following general steps for adsorption processes: (1) diffusion of lixiviant molecules through the liquid film layer, (2) diffusion of lixiviant through the particle or product layer, (3) adsorption of the lixiviant to the particle surface, (4) surface reaction between the metal and lixiviant to form product, (5) desorption of the product from the particle surface, (6) diffusion of the product from within the particle or through the product layer, and (7) diffusion of product back through the liquid boundary layer. It is also possible for REEs to directly react with lixiviant at the outer surface of the particle without diffusion through the particle or product layer, thus omitting steps 2 and 6. In the SCM, metals embedded within solid particles react with lixiviant to form a product layer on the particle surface through which diffusion occurs (Figure 1) [22]. In a well-mixed system, the rate limiting mechanism is controlled by film diffusion (steps 1 and 7), product layer diffusion (steps 2 and 6), or the chemical reaction (step 4) [23]. Each type of mechanism corresponds to an equation that produces a straight line when plotted versus time (Table 1). The model that gives the best fit line represents the rate limiting step.

Table 1. Select geometric contraction models fit to MKT leaching data to describe the extraction process of cationic species. The shrinking core model (SCM) is used to determine the physical phenomenon responsible for limiting the rate of species extraction from the ore. SCM is composed of three parts: liquid film diffusion, chemical reaction, and product layer diffusion. The Jander equation (JE) and Ginstling–Brounshtein model (GBM) both describe leaching processes which are diffusion limited. Each equation results in the rate constant (k) to describe the kinetics of the reaction.

Geometric Contraction Model	Equation
Shrinking Core Model	
Liquid Film Diffusion Control	$X = \frac{M_t}{M_i} = kt$
Product Layer Diffusion Control	$1 - 3(1 - X)^{\frac{2}{3}} + 2(1 - X) = kt$
Chemical Reaction Control	$1 - (1 - X)^{\frac{1}{3}} = kt$
Jander Equation	$\left[1 - (1 - X)^{\frac{1}{3}}\right]^2 = kt$
Ginstling–Brounshtein Model	$1 - \frac{2}{3}X - (1 - X)^{\frac{2}{3}} = kt$

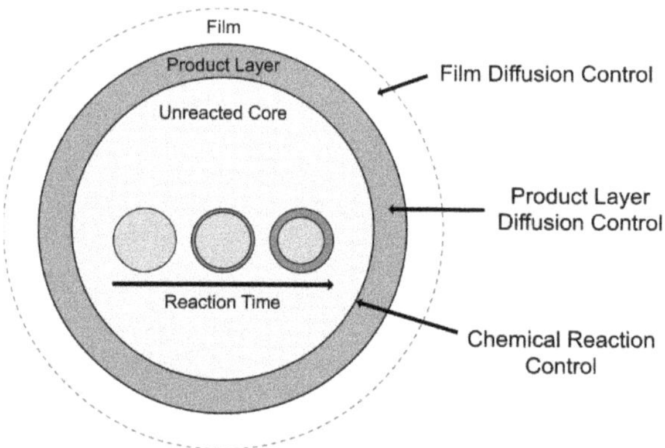

Figure 1. Depiction of the shrinking core model in which the unreacted core is converted to a product layer. The rate limiting mechanism may follow film diffusion control, product layer diffusion control, or chemical reaction control at the particle surface.

After determining that the leaching mechanism was potentially a diffusion-limited process, the data were then fit to JE and the GBM to validate this claim [24,25]. JE is the simplest one-dimensional diffusion control model for a spherical particle. It is a valid approximation of diffusion behavior at low conversions because it assumes the reacting particles maintain a constant volume and interfacial area so that two- and three-dimensional diffusions are treated as one-dimensional [26]. The GBM, on the other hand, can apply to systems where the solid maintains a relatively constant volume, but where interfacial reaction area changes over time. Like the SCM, the overall volume of the particle remains the same because the resulting product layer replaces the reacted volume. The GBM is similar to the product layer diffusion equation where kt is multiplied by a factor of 3.

To overcome potential underestimations in mass of the leached species, the rate limiting mechanism was determined by evaluating the leaching process in terms of the heterogeneous sorption between REEs and citrate rather than the mass leached from the underclay. To do this, the leachate data were fitted to a modified diffusion control model (MDCM) based on the intraparticle diffusion model introduced by Crank (1975), which describes the sorption process in terms of Fick's law for non-stationary spherical particles in a homogeneous solvent (Equation (1)) [27]. The Crank model uses the fraction of species leached into the solution (F), which is determined by measuring the concentration of the species in the leachate at time t (q_t) and at equilibrium (q_e), to determine the diffusion rate constant (B). Once B is determined, the apparent intraparticle diffusion coefficient (D_e) is calculated (Equation (2)). For this study, MDCM was applied to estimate q_e as well as the amount of REE leaching by non-diffusive behavior (q_0) by fitting the measured q_t versus time (Equation (3)) [28]. For both models, n represents the number of iterations used to obtain the best fit line. In this case, summation from n = 1 to n = 6 yielded the best fit, with all R-squared values greater than 0.99.

$$F = \frac{q_t}{q_e} = 1 - \frac{6}{\pi^2}\sum_{n=1}^{\infty} \frac{1}{n^2} e^{-n^2 Bt} \tag{1}$$

$$B = \frac{\pi^2 D_e}{r^2} \tag{2}$$

$$q_t = q_e\left[1 - \left(1 - \frac{q_0}{q_e}\right)\frac{6}{\pi^2}\sum_{n=1}^{\infty} e^{-n^2 B(t-t_0)}\right] \tag{3}$$

The kinetic models in eqns. 1 and 3 are used to describe diffusion-controlled processes, but they do not distinguish between the types of diffusion which limit the rate of reaction. To determine the type of diffusion control, Boyd plots were used to determine if the rate is limited by intraparticle or film diffusion [29,30]. This was carried out by plotting the Bt term vs. t; where Bt is calculated by one of two equations depending on whether F is >0.85 or ≤0.85 (Equations (4)–(6)) [31,32]. If the results of the Boyd plot pass through the origin, the mechanism is considered to be intraparticle diffusion controlled, otherwise film diffusion is the dominant mechanism. For this study, y-intercept values < |0.1| are considered to pass through the origin.

$$F = 1 - \frac{6}{\pi^2} e^{-Bt} \tag{4}$$

$$Bt = -\ln\left(\frac{\pi^2}{6}\right) - \ln(1 - F(t));\ F(t) > 0.85 \tag{5}$$

$$Bt = \left(\sqrt{\pi} - \sqrt{\pi - \frac{\pi^2 F(t)}{3}}\right)^2;\ F(t) \le 0.85 \tag{6}$$

After the rate limiting mechanism is determined, an empirical kinetic model can be applied to calculate the rate constant and rate of change. The two kinetic equations chosen to fit to the leaching data are the Weber–Morris intraparticle diffusion model (IPD) (Equation (7)) and the general power law equation (Equation (8)), both of which are used to describe sorption kinetics [33,34]. To fit these models, only data for the first 96 hr were used to capture the leaching kinetics prior to reaching an equilibrium state. IPD provides a good representation of Fickian diffusion, but it is based on the geometry of a plate [35]. Therefore, it is likely not an appropriate model considering the spherical geometry of the particles. Instead, the power law can be applied as an empirical model to provide a more accurate fit to the particulate dataset. Similar to IPD, the power law is a hyperbolic function but with a variable exponent that describes the release rate of the species into solution (Equation (8)). When n < 1, the rate of release decreases over time whereas n > 1 signifies an increase in release rate. Additionally, a value of n = 1 means that the rate is constant over the course of the experiment and yields a straight-line slope. The farther n is from 1 signifies a faster decrease (n < 1) or faster increase (n > 1). Small changes in rate occur at values near 1. In both models, k represents the rate constant for each respective model and C is a constant that quantifies the amount of deviation from the model.

$$q_t = k\sqrt{t} + C \tag{7}$$

$$q_t = kt^n \tag{8}$$

3. Results and Discussion

3.1. Extraction Efficiencies

Montross et al. (2020) demonstrated extraction efficiencies of 30% total REE underclay using 0.1 M citric acid–sodium citrate buffer solution amended with 0.5 M NaCl at pH 5.8. In comparison, the leaching of REEs by 0.1 M citric acid–sodium citrate in scaled-up experiments only reached a TREE extraction efficiency of 11% over 18 days as reported in Table 2. There are several factors that may contribute to the lower extraction efficiencies achieved in these samples, including the pH, concentration of exchangeable cations, and the leaching method. Citric acid is a polyprotic organic acid with pKa values of 3.13, 4.76, and 6.40 [36]. At high pH values, citric acid becomes deprotonated and exists mainly as anionic Cit^{3-} [37]. Lower pH values are expected to result in increased extraction efficiencies because the acidic protons exchange with REEs. The fact that this was not the case in the scaled-up experiments suggests that pH was not the dominant factor affecting the leaching process when gangue elements are not quickly dissolved by the acid to release REEs into the solution. Because citric acid chelates with charged species at the surface of IACs, it is possible that these sources make up a smaller fraction of feedstock material, resulting in overall lower amounts of extractable REEs.

Table 2. REE and cation leaching efficiencies (wt.%) after 430 hr using 0.1 M citric acid–citrate leaching solution as modified from Yang et al. (2022) [16].

REE	Wt. %	Major Cations	Wt. %
Sc	7.15	Mg	4.03
Y	18.9	K	0.578
La	1.06	Ca	31.0
Ce	3.08	Al	0.275
Pr	5.18	Si	0.146
Nd	8.56	P	26.0
Sm	23.9	Mn	3.26
Eu	34.1	Fe	2.83
Gd	36.6	Zn	24.2
Tb	34.2		
Dy	27.0	U	8.61
Ho	20.0	Th	25.0
Er	15.2		
Tm	11.4	Cr	0.387
Yb	9.64	Co	28.5
Lu	8.71	Ni	15.3
TREE	10.8	Cu	13.9

The feedstock material has a large impact on the overall leaching efficiency. Multi-modal characterization of these underclays determined the presence of REE–phosphates (e.g., monazite and xenotime or their hydrated forms of rhabdophane and churchite), apatitic phases, and degradation products such as crandallite embedded within the underclays [14]. Citric acid is an effective organic ligand in REE complexation of phosphate minerals [38] and increases the solubility of P [39]. Apatite, for example, is dissolved by ionic detachment of calcium and orthophosphate ions from the mineral surface. The high extraction efficiencies of Ca and P suggest a significant amount of the phosphate and apatitic phases are solubilized by the citric acid lixiviant (Table 2). Other elements like Al and Si are abundant within underclays and can be indicative of the dissolution of IACs in a citric acid environment [40]. As shown in Table 2, less than 1% of Al and Si available in the underclay samples is extracted into the leachate over 18 days. If the releases of Al and Si are indicative of the extent of IAC dissolution, low Al and Si extraction efficiencies suggest little breakdown of IAC phases, especially compared to phases primarily containing Ca and P. Therefore, the REEs released from IAC phases are more likely to come from the easily accessible surface-bound REEs rather than REEs embedded within the underclay matrix. The dissolution of P and Ca, on the other hand, suggests REEs may be released from within phosphate and apatitic phases to a greater extent than from within IACs.

The distribution of gangue elements in the leachate solution suggests there is a greater breakdown of phosphate and apatitic phases compared to IACs. This is supported by the distribution of REEs extracted into the leachate which shows greater extraction efficiencies for middle REEs (MREEs) including Sm, Eu, and Gd. The strong MREE enrichment of the leachate solution suggests that calcium- and phosphate-bearing minerals could be a significant source of the REE extracted from these samples. Citric acid has been shown to primarily extract MREEs from apatite phases, further adding evidence for this claim [38]. Meanwhile, Y and the heavy REEs (HREEs) Tb, Dy, Ho, Er, Tm, Yb, and Lu are more likely to be leached from the surfaces of the clay matrix or mineral surfaces [14]. The amounts of HREEs extracted into the leachate are much lower than those of MREEs, suggesting that a smaller portion of the REEs are from IAC sources. Based on the particularly low extraction efficiencies of Al and Si, very little IAC may be dissolved which may limit the amount of REEs that can be released from this material, especially when compared to phosphate-bearing minerals. Several critical minerals (e.g., Co, Ni, and Cu) also show higher extraction efficiencies than most of the other cations present in the leachate. In particular, 28.5% of Co is extracted by the end of the experiment, on par with the efficiencies obtained for Ca, P, and MREEs (Table 2). The extraction efficiencies of these CMs are much higher than that of Fe and Mn but lower than that of Zn, all of which come from the same d-block elements. This suggests that the elemental group may not be the dominant factor contributing to extractability. Instead, the feedstock source may play a larger role in the rate of REE extraction and ultimately the leaching efficiency.

3.2. Leaching Profiles

The dissolution and precipitation of metals in solution typically follows one of several kinetic profiles based on the time-dependent leaching behavior [41,42]. As seen in Figure 2, the TREE leaching profile shows two distinct kinetic regimes—an initial leaching phase followed by a second slower phase until equilibrium is reached. The presence of two stages reveals a shift in leaching rate due to changes in the rate limiting mechanism or exhaustion of the species from a specific source. Given the presence of IAC phases in the MKT underclay samples [14], the initial increase in concentration may be dominated by the leaching of ion-exchangeable REEs from the easily accessible IAC surfaces. Once these sources of REEs are exhausted, the mechanism may transition to a diffusion-controlled process from within less accessible REEs associated with the clay and mineral phases.

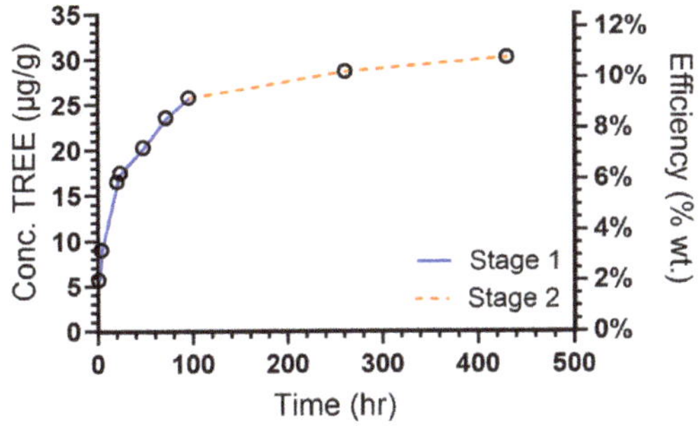

Figure 2. Kinetic profile of TREEs leached by 0.1 M citric acid at ambient temperature and pressure shows two distinct kinetic regimes.

Cations such as Ca and P share similar kinetic behavior to REEs, indicating that they follow a similar leaching mechanism or are leached from a similar source material (Figure 3a). As mentioned previously, Ca and P are major cations in phosphate-bearing minerals such as apatite. Their release into solution at such extraction efficiencies suggests a dissolution of the apatite minerals which may cause the exposure and subsequent leaching of the embedded REEs. However, this is not the case for other gangue elements like Al, Si, and Fe whose concentrations continue to rise over the course of the 18-day experiment (Figure 3b). These elements exhibit a different overall leaching mechanism and more closely follow a hyperbolic fashion. The absence of a plateau indicates that the reaction kinetics for these gangue elements have not reached equilibrium by the conclusion of the mixing barrel experiments. This suggests that much slower leaching rates occur in gangue elements when compared to TREEs. The leaching profiles of critical minerals is represented in Figure 3c by the elements Co, Ni, and Cu. Like the TREE profile (Figure 3a), CM concentration in the leachate reaches an equilibrium within the timeframe of the experiments. The steepness in the leaching profiles of CM suggests that the majority of the leachable species are extracted within the 18-day experiment. This is in contrast to the leaching profile of Fe (Figure 3b), despite all being transition metals. The difference in leaching profile between Fe and other CMs may be influenced by the accessibility of the element within the feedstock material.

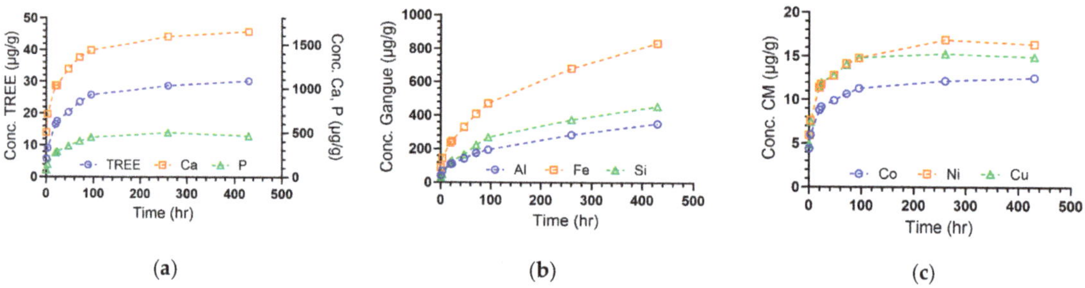

Figure 3. Leaching profiles of (**a**) TREEs, Ca, and P, (**b**) gangue elements Al, Fe, and Si, and (**c**) critical minerals Co, Ni, Cu.

3.3. *Rate Determining Mechanism*

Each of the equations summarized in Table 2 is based on the fractional conversion (X) of the solid which is the reacted mass divided by the initial mass of the reactant ($X = M_t/M_i$). X is typically calculated by weighing the solid before and after the reaction to determine the total mass that is released from the solid particles. The concentration data retrieved from the mixing barrel experiments correspond to the amount of solubilized REEs leached into the solution from the solid particles. Applying the SCM, JE, and GBM to the mixing barrel data, all displayed good fits with JE having the best fit for the first 4 days (stage 1) and SCM film diffusion fitting days 4–18 (stage 2) best (Figure 4). However, many fits were nearly as good, causing ambiguity over which equation represents the rate limiting mechanism. For instance, both film diffusion and product layer diffusion are nearly identical in fit during the second stage of the process, indicating that diffusion may have mixed control mechanisms in this regime.

3.4. *Modified Diffusion-Controlled Mechanism*

While the geometrical contraction models provide good fits to the MKT data, the masses used to fit these models do not account for the total mass of the particle that is released from the underclay through complexation with the lixiviant. Rather, MDCM (Equation (3)) is a better fit to the leachate data ($R^2 = 0.997$) and fits the entire dataset non-linearly rather than fitting two linear equations to the different kinetic regimes. The good fit of the TREE concentration data to MCDM validates that the main source of REEs from these samples undergoes diffusion control as the rate limiting mechanism (Figure 5). The fit also reveals very slow diffusion with an apparent diffusion coefficient of 2.86×10^{-9} cm^2 s^{-1}, which was calculated from Equation (2). For comparison, the diffusion coefficient of REEs out of mineral phases at 600–800 °C is about 2×10^{-6} cm^2 s^{-1} [43]. The coefficients for the gangue elements show relative rates of diffusion with underclay-associated cations having the slowest rates of diffusion, such as Al and Si having diffusion coefficients of 4.28×10^{-10} cm^2 s^{-1} and 2.57×10^{-10} cm^2 s^{-1}, respectively (Table 3). Meanwhile, cations associated with phosphate-bearing minerals have coefficients an order of magnitude higher than those associated with clays. This suggests that the dissolution of apatite material occurs to a higher degree than in clay minerals at pH 2.8.

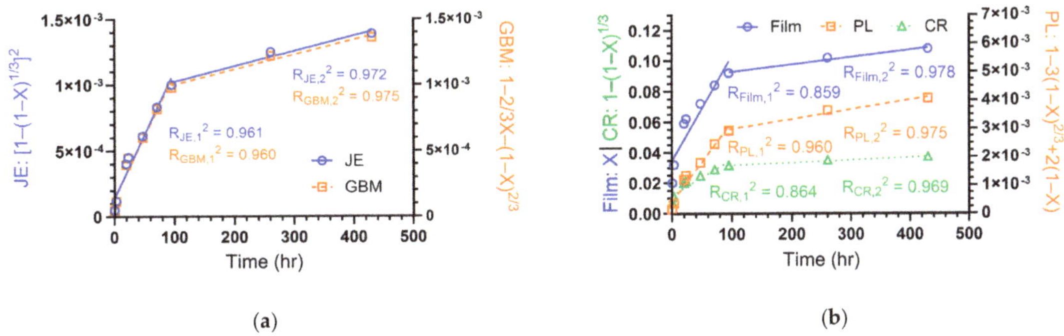

Figure 4. TREE concentration data fits to (**a**) Jander equation (JE) and Ginstling–Brounshtein model (GBM) and (**b**) film diffusion, product layer diffusion (PL), and chemical reaction (CR) control models.

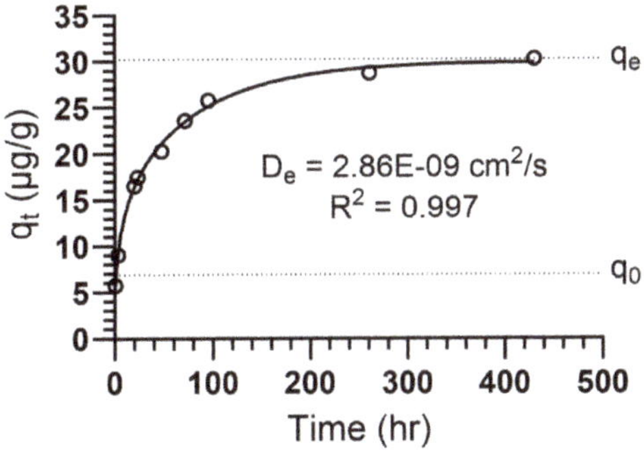

Figure 5. TREE leaching data fitted to the modified diffusion control model (MDCM) (Equation (3)).

In addition to the diffusion coefficients and rate constants, the MDCM is useful for estimating the equilibrium concentration of REEs in the leachate (q_e) as well as the concentration of species leached from the underclay by non-diffusion-controlled mechanisms (q_0). Table 3 reveals that a measurable portion of REEs is leached into the solution by non-diffusive means. This amount of REEs is represented by q_0 as a percentage of q_e, which for TREEs is 22.2%. This means that over 20% of the TREEs leached into solution come from non-diffusion-limited mechanisms, such as chemical reaction or surface adsorption controls. The greater presence of HREEs that are leached by non-diffusive mechanisms is expected since HREEs tend to leach from the surfaces of IAC and mineral phases. Interestingly, this was also the percentage of ion-exchangeable REEs measured by leaching with ammonium sulfate by Montross et al. (2020) compared to the total amount of REEs leached by citric acid (10 μg/g ion-exchangeable REEs using $(NH_4)_2SO_4$ and 45 μg/g total REEs using citrate) [16]. These results validate the MDCM as a method for elucidating the rate controlling mechanisms to measure REE leachate from underclay feedstocks using weak organic acid.

Table 3. Parameters from the modified diffusion control model (MDCM) for individual and total REEs and relevant cations. The correlated output parameters include q_e, q_0, B, and t_0, while D_e and % non-diffusive were calculated from these values.

REE	q_e (µg g^{-1})	q_0 (µg g^{-1})	B (hr^{-1})	t_0 (hr)	R-Squared	D_e (cm^2 s^{-1})	Non-Diffusive (%)
Sc	1.79	2.52×10^{-1}	5.70×10^{-3}	5.60	0.996	1.38×10^{-9}	14.1
Y	10.5	2.98	1.20×10^{-2}	3.86	0.998	2.91×10^{-9}	28.2
La	4.24×10^{-1}	0	1.08×10^{-2}	0	0.996	2.61×10^{-9}	0
Ce	2.57	1.55×10^{-1}	1.12×10^{-2}	0.965	0.995	2.72×10^{-9}	6.01
Pr	5.24×10^{-1}	5.76×10^{-2}	1.12×10^{-2}	1.96	0.996	2.72×10^{-9}	11.0
Nd	3.48	4.41×10^{-1}	1.16×10^{-2}	2.07	0.996	2.81×10^{-9}	12.7
Sm	2.17	4.57×10^{-1}	1.14×10^{-2}	3.38	0.996	2.78×10^{-9}	21.0
Eu	6.79×10^{-1}	7.96×10^{-2}	1.47×10^{-2}	1.14	0.994	3.58×10^{-9}	11.7
Gd	4.42	9.84×10^{-1}	1.23×10^{-2}	3.12	0.997	2.98×10^{-9}	22.3
Tb	6.48×10^{-1}	1.79×10^{-1}	1.10×10^{-2}	4.49	0.997	2.67×10^{-9}	27.6
Dy	2.78	7.74×10^{-1}	1.14×10^{-2}	4.20	0.997	2.77×10^{-9}	27.8
Ho	3.96×10^{-1}	9.12×10^{-2}	1.35×10^{-2}	2.65	0.997	3.29×10^{-9}	23.0
Er	8.00×10^{-1}	2.47×10^{-1}	1.16×10^{-2}	4.52	0.997	2.81×10^{-9}	30.8
Tm	8.10×10^{-2}	2.42×10^{-2}	9.98×10^{-2}	4.95	0.998	2.42×10^{-9}	29.9
Yb	4.01×10^{-1}	9.58×10^{-2}	1.13×10^{-2}	3.09	0.997	2.74×10^{-9}	23.9
Lu	5.19×10^{-2}	1.48×10^{-2}	1.10×10^{-2}	4.23	0.996	2.68×10^{-9}	28.4
TREE	30.0	6.65	1.18×10^{-2}	3.23	0.997	2.86×10^{-9}	22.2
Major Cations	**q_e (µg g^{-1})**	**q_0 (µg g^{-1})**	**B (hr^{-1})**	**t_0 (hr)**	**R-Squared**	**D_e (cm^2 s^{-1})**	**Non-Diffusive (%)**
Mg	227	105	3.97×10^{-3}	15.4	0.985	9.64×10^{-10}	46.2
K	143	80.0	1.09×10^{-2}	5.78	0.974	2.64×10^{-9}	56.0
Ca	1640	591	1.16×10^{-2}	3.69	0.998	2.81×10^{-9}	36.0
Al	483	17.6	1.76×10^{-3}	3.95	0.996	4.28×10^{-10}	3.63
Si	716	104	1.06×10^{-3}	38.3	0.997	2.57×10^{-10}	14.5
P	486	22.4	1.78×10^{-2}	0.302	0.994	4.32×10^{-9}	4.61
Mn	5.74	2.08	3.60×10^{-3}	14.0	0.992	8.74×10^{-10}	36.2
Fe	1170	39.1	1.72×10^{-3}	6.11	0.998	4.17×10^{-10}	3.34
Zn	31.3	6.79	6.21×10^{-3}	5.50	0.995	1.51×10^{-9}	21.7
U	5.15×10^{-1}	1.05×10^{-1}	1.07×10^{-2}	2.22	0.998	2.61×10^{-9}	20.4
Th	4.25	3.41×10^{-1}	1.20×10^{-2}	1.08	0.999	2.92×10^{-9}	8.01
Ti	3.79	0	9.73×10^{-4}	7.86	0.994	2.36×10^{-10}	0
Cr	1.64	0	3.31×10^{-4}	57.4	0.994	8.04×10^{-11}	0
Co	12.6	6.90	9.84×10^{-3}	7.92	0.998	2.39×10^{-9}	54.8
Ni	16.7	7.62	1.16×10^{-2}	5.05	0.995	2.81×10^{-9}	45.5
Cu	15.1	6.37	2.36×10^{-2}	2.81	0.995	5.73×10^{-9}	42.1

A comparison of major cations shows that Ca, which is typically associated with mineral phases, has 36.0% of its extractable content leached by non-diffusive mechanisms (Table 3). Meanwhile, P only has 4.61% of its leached content attributed to non-diffusive means. This discrepancy could arise because apatite tends to dissolve by the ionic detachment of Ca cations. Meanwhile, low molecular weight organic acids like citrate complex with the P on the material [39]. This could expose more surfaces where Ca is available to be leached, resulting in higher extraction efficiencies than P. Underclay-associated elements Al and Si also show contrasting results to one another (Table 3). Only 3.63% of Al is leached by non-diffusive means which is expected since it is embedded in the underclay matrix. Si, however, leaches 14.5% of its content by the non-diffusive controls. Much of the Al and Si in the mixing barrel samples are expected to come from underclay sources such as illite [14]. The structure of illite consists of two layers of silica tetrahedrons surrounding a single layer of alumina octahedrons [40]. Hence, the Si from the outer layers of silica may be more easily leached without undergoing diffusion through the underclay matrix. Since very small percentages of both Si and Al are leached from within solid particles, this leads to more Si being leached into the solution from the outer surfaces of the IAC.

Ni, Co, and Cu were also leached into the solution by non-diffusive means within the first 2–10 hr of the leaching process (Table 3). The fact that over 40% of these species (over 50% for Co) are leached by non-diffusive means indicates that they are easily leachable from the feedstock material. This could be beneficial for leaching these species in short time frames. However, the overall low extraction efficiencies presented in Table 1 reach a plateau earlier than the other cations measured. This suggests that while much of these CMs are leached from the mineral quickly, a significant portion is extracted by non-diffusive means and small percentage of the overall CM content available in the original sample is leached by diffusion from within the clay particles.

3.5. Determination of the Diffusion-Controlled Mechanism

The MDCM (Equation (3)) describes when a process is diffusion controlled but it does not distinguish the types of diffusion that may be occurring in the system. Therefore, Boyd plots were used to distinguish the extraction process as film or intraparticle diffusion controlled. The results of the Boyd plot show the TREE data pass through the origin (y_{int} = 0.0502), indicating that the leaching mechanism is controlled by intraparticle diffusion (Figure 6a). Most other cations also exhibited intraparticle diffusion as the rate limiting step, including Al, Si, and P. The Boyd plot for Ca, however, did not pass through the origin which suggests that film diffusion is the rate limiting mechanism for Ca (Figure 6b). During the first 96 hr (prior to equilibrium), the resulting line does not pass through the origin (y_{int} = 0.163). This observation signifies that film diffusion plays a large role in the Ca leaching mechanism over the entire course of the experiments rather than intraparticle diffusion. As mentioned previously, Ca ions embedded in the surface of the apatite minerals readily chelate with citric acid and leach into the solution. Therefore, they are likely more accessible for extraction from the surface of the clay particles. Other species share a similar trend with Ca and are dominated by film diffusion. These elements include the cations Mg, Mn, K, Ni, Co, and Cu (Table 4). These elements also show >30% of their content leached by non-diffusive means (Table 3), further indicating that a surface-controlled process such as film diffusion most likely controls the leaching rate of these species.

Table 4. Y-intercept values obtained from Boyd plots. Values < |0.1| signify species in which diffusion is dominated by intraparticle diffusion. Otherwise, it undergoes a film diffusion-controlled mechanism.

REE	Y-Int	R-Squared	Major Cations	Y-Int	R-Squared
Sc	6.58×10^{-3}	0.984	Mg	1.74×10^{-1}	0.938
Y	8.77×10^{-2}	0.992	K	3.41×10^{-1}	0.927
La	-2.84×10^{-2}	0.981	Ca	1.63×10^{-1}	0.992
Ce	-1.04×10^{-2}	0.985	Al	8.80×10^{-3}	0.996
Pr	9.72×10^{-4}	0.987	Si	-5.65×10^{-3}	0.975
Nd	5.15×10^{-3}	0.985	P	-2.45×10^{-2}	0.982
Sm	3.50×10^{-2}	0.990	Mn	1.01×10^{-1}	0.968
Eu	1.86×10^{-2}	0.981	Fe	2.56×10^{-3}	0.995
Gd	4.97×10^{-2}	0.992	Zn	4.00×10^{-2}	0.988
Tb	6.45×10^{-2}	0.987			
Dy	7.31×10^{-2}	0.992	U	6.14×10^{-2}	0.991
Ho	5.99×10^{-2}	0.992	Th	1.16×10^{-2}	0.995
Er	8.37×10^{-2}	0.986	Ti	4.23×10^{-4}	0.966
Tm	9.01×10^{-2}	0.987	Cr	-1.52×10^{-3}	0.955
Yb	6.54×10^{-2}	0.994	Co	2.77×10^{-1}	0.961
Lu	8.47×10^{-2}	0.982	Ni	2.51×10^{-1}	0.968
TREE	5.02×10^{-2}	0.992	Cu	1.81×10^{-1}	0.967

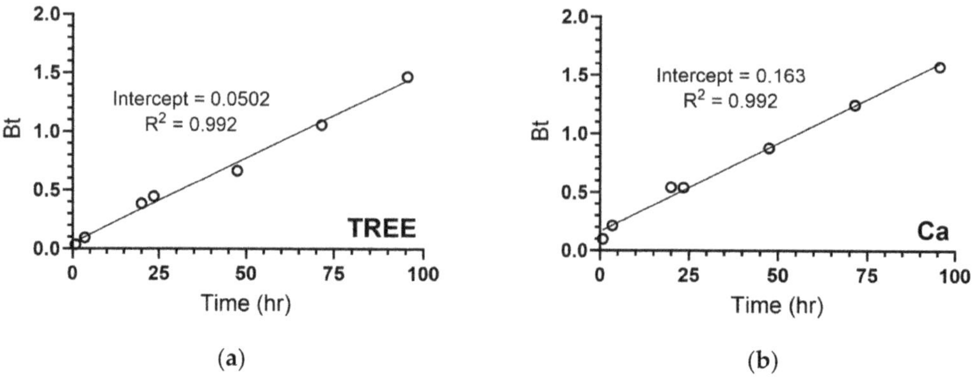

Figure 6. Boyd plots of the (**a**) TREE and (**b**) Ca cation leaching process.

3.6. Reaction Kinetics

Now that it is established that the rate limiting mechanism of the REE leaching process is intraparticle diffusion controlled, the rate can be described by a kinetic equation corresponding to intraparticle diffusion. Plotting the TREE concentration data to the generic power law model provided a better fit than the IPD model with R-squared values of 0.996 and 0.976, respectively (Figure 7). The n value of the power law is 0.305 which is farther away from the value of 1 than the 0.5 exponent used in IPD. This signifies that the rate of REE extraction decreases faster than what is modeled by the IPD equation. Consistent with this claim, the power law parameters of REE and the major cations tabulated in Table 5 show that the elements with the slowest rate of decrease have the highest values for n (but less than 1). The kinetic rate constant also provides a measure of how fast the process occurs with higher values meaning that the element is leached into solution at a faster rate. However, these values cannot be compared to one another because the units differ depending on the value of n for the element of interest. Underclay-associated elements such as Al and Si have higher values of n and therefore diffuse from the clay or minerals with a slower rate of decrease than REEs. This is consistent with the fact that the leachate concentrations of these elements do not reach equilibrium during the course of the mixing barrel experiments.

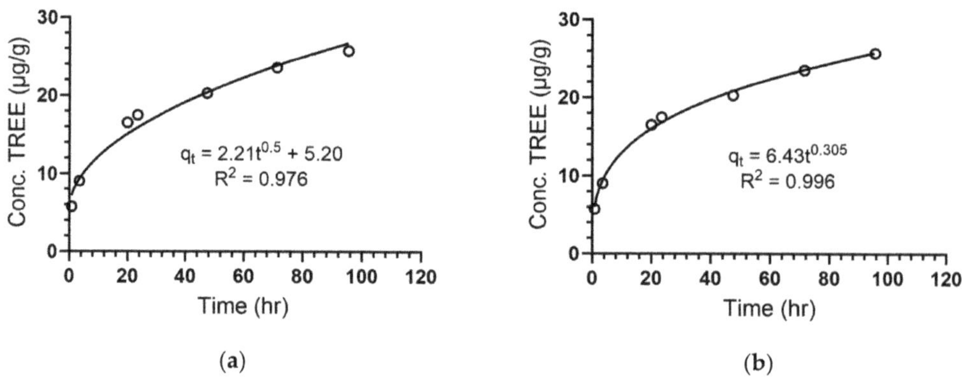

Figure 7. TREE concentration data fitted to (**a**) the Weber–Morris intraparticle diffusion model (IPD) and (**b**) the power law model.

Table 5. Power law model parameters for individual and total REEs and major cations influenced by intraparticle diffusion behavior (excludes species controlled by film diffusion).

REE	k ($\mu g\ g^{-1}\ h^{-n}$)	n	R-Squared	Major Cations	k ($\mu g\ g^{-1}\ h^{-n}$)	n	R-Squared
Sc	2.13×10^{-1}	0.385	0.990	Al	39.3	0.345	0.989
Y	2.58	0.280	0.996	Si	23.5	0.527	0.991
La	5.38×10^{-2}	0.405	0.997	P	93.2	0.345	0.997
Ce	3.88×10^{-1}	0.372	0.996	Fe	75.5	0.393	0.986
Pr	8.54×10^{-2}	0.357	0.997	Zn	5.62	0.305	0.993
Nd	6.10×10^{-1}	0.347	0.996				
Sm	4.35×10^{-1}	0.319	0.996	U	1.18×10^{-1}	0.282	0.997
Eu	1.37×10^{-1}	0.326	0.991	Th	7.15×10^{-1}	0.352	0.998
Gd	9.61×10^{-1}	0.304	0.996				
Tb	1.43×10^{-1}	0.299	0.993	Ti	1.62×10^{-1}	0.413	0.970
Dy	6.43×10^{-1}	0.290	0.995	Cr	7.18×10^{-3}	0.762	0.980
Ho	9.44×10^{-2}	0.289	0.997				
Er	1.96×10^{-1}	0.279	0.993				
Tm	1.92×10^{-2}	0.278	0.995				
Yb	9.34×10^{-2}	0.284	0.998				
Lu	1.24×10^{-2}	0.282	0.992				
TREE	6.43	0.305	0.996				

4. Conclusions

The Jander equation, shrinking core, and Ginstling–Brounshtein kinetic models were examined to determine the kinetic profiles and type of leaching behavior of REEs and other major cations using experimental data for a Middle Kittanning underclay that was subjected to a citric acid solution. However, none of these models accurately describes the full set of experimental leachate concentration data as it is necessary to break the full reaction time down into two distinct kinetic regimes. A modified diffusion-controlled adsorption kinetics model, which also accounts for a relatively small fraction of species with kinetics governed by non-diffusion-controlled mechanisms, was shown to best fit the full set of experimental leachate concentration data. An initial rapid leaching of surface REEs is followed by a slow intraparticle diffusion mechanism. The leaching profiles of REEs and gangue cations are comparable to Ca and P with similar kinetic profiles and rate parameters. A greater percentage of MREEs were extracted from the solid particles than HREEs or LREEs, further suggesting REEs are primarily leached from the phosphate-bearing phases rather than the underclay material itself. Critical minerals (e.g., Co, Ni, and Cu) follow a similar trend as REEs at a faster initial leaching rate. Other cations associated with the underclay materials such as Al and Si exhibited much slower kinetics and did not reach equilibrium in the experimental time frame. This is likely due to the underclay material being primarily composed of a phyllosilicate matrix which is insoluble and may not undergo the same leaching mechanisms.

The primary leaching mechanism over the 18-day experiment was determined to be diffusion controlled. Using the modified diffusion-control model, the apparent diffusion coefficient was calculated to be 2.86×10^{-9} $cm^2\ s^{-1}$ for TREEs. This was on par with the diffusion parameters of Ca and P but an order of magnitude greater than that of Al and Si. This strongly suggests that REEs are predominately leached from the same phases as Ca and P and most likely come from mineral surfaces (e.g., hydroxyapatite or crandallite) rather than diffusion from the clay matrix. In addition, the diffusion coefficients of CMs are on the same order of magnitude as those of REEs. The MDCM model indicated that up to 22% of the TREEs leached into solution occurred by non-diffusive mechanisms, which represents the amount of REEs leached directly from the particle surfaces in the initial hours of the leaching process. This number is similar to the ion-adsorbed REEs measured by Montross et al. (2020) using ammonium sulfate, signifying this may be the main source of the non-diffusive REEs. It is important to note that it is not clear if these surface-reacted REEs come from ion-adsorbed phases or subsequent mineral dissolution as new surfaces are exposed.

Boyd plots were employed to determine the leaching mechanism of REEs after establishing the kinetic mechanism as diffusion controlled. Those results indicated that the REEs were leached primarily as intraparticle diffusion rather than film control. However, Ca and the CMs Ni, Co, and Cu also appear to follow film diffusion control mechanisms even though the Boyd plots indicate that the kinetic mechanism for recovery of these species is primarily controlled by film diffusion. A proposed mechanism is that Ca, P, and easily accessible REE species are first leached from the particle surface of apatitic phases within the first several hours of the extraction. Once these species are leached, the mechanism for the REEs and P shifts to intraparticle diffusion control for the remainder of the reaction time. This behavior suggests that the majority of REEs which are extracted into the leachate come from within the particle, either from specific REE-bearing minerals or within the underclay matrix.

The gangue cations (Al and Si) have less non-diffusive leaching than REE species. It is likely that these elements are dominated by diffusion with strong ionization energy based on the low extraction efficiencies and slow kinetics. Further investigation is needed to confirm whether the diffusion limits observed are a result of inherent intraparticle diffusion, product layer formation, or both. Based on our models, it is likely a combination of both, but additional imaging techniques are necessary to verify this hypothesis. We recognize there is additional work to be carried out to examine the precipitation of rare earth oxalates similar to the preliminary findings in Yang et al. (2022) as well as their influence on REE trace element extraction [16].

The modeling results from this study coupled with the experimental findings from previous benchtop and upscaled experiments show the importance of experimental design parameters on the leaching kinetics and overall extraction efficiencies of REEs and major cations from underclay materials. Taken as a whole, these findings are necessary for the development of technoeconimic analysis to evaluate the economic performance of citric acid leaching of REEs under various conditions and inform strategies for domestic and international recovery of REEs and CMs to implement sustainable efforts to make use of underutilized coal waste streams and adjacent IAC deposits.

Author Contributions: Conceptualization, W.B., J.Y. and C.V.; methodology, P.P. and W.B.; software, P.P.; validation, P.P, W.B., J.Y. and C.V.; formal analysis, P.P.; investigation, P.P.; data curation, J.Y. and C.V.; writing—original draft preparation, P.P. and C.V.; writing—review and editing, W.B., J.Y. and C.V.; supervision, W.B. and C.V.; project administration, C.V.; funding acquisition, C.V. All authors have read and agreed to the published version of the manuscript.

Funding: This work was performed in support of the U.S. Department of Energy's (DOE) Fossil Energy and Carbon Management, Rare Earth Elements/Minerals Sustainability Field Work Proposal 1022420.

Data Availability Statement: Data supporting reported results can be found at https://edx.netl.doe.gov/ree-cm/.

Acknowledgments: This work was carried out under the ORISE Post-graduate Research Program at NETL. The authors wish to thank Burt Thomas (Technical Portfolio Lead) for programmatic guidance, direction, and support. Priscilla Prem would also like to thank the Mickey Leland Energy Fellowship Program for initiating her participation in this project.

Conflicts of Interest: The authors declare no conflict of interest.

References

1. Haque, N.; Hughes, A.; Lim, S.; Vernon, C. Rare Earth Elements: Overview of Mining, Mineralogy, Uses, Sustainability and Environmental Impact. *Resources* **2014**, *3*, 614–635. [CrossRef]
2. Imholte, D.D.; Nguyen, R.T.; Vedantam, A.; Brown, M.; Iyer, A.; Smith, B.J.; Collins, J.W.; Anderson, C.G.; O'Kelley, B. An assessment of U.S. rare earth availability for supporting U.S. wind energy growth targets. *Energy Policy* **2018**, *113*, 294–305. [CrossRef]
3. U.S. Geological Survery (USGS). Rare Earths Mineral Commodity Summaries. 2021. Available online: https://pubs.usgs.gov/periodicals/mcs2021/mcs2021-rare-earths.pdf (accessed on 4 August 2021).
4. Yang, X.J.; Lin, A.; Li, X.-L.; Wu, Y.; Zhou, W.; Chen, Z. China's ion-adsorption rare earth resources, mining consequences and preservation. *Environ. Dev.* **2013**, *8*, 131–136. [CrossRef]
5. Kanazawa, Y.; Kamitani, M. Rare earth minerals and resources in the world. *J. Alloys Compd.* **2006**, *408–412*, 1339–1343. [CrossRef]
6. Papangelakis, V.G.; Moldoveanu, G. In *Recovery of Rare Earth Elements from Clay Minerals, Proceedings of the 1st Rare Earth Resources Conference, Milos, Greece, 4–9 June 2014*. pp. 191–202. Available online: https://www.eurare.org/docs/eres2014/fifthSession/VladimirosPapangelakis.pdf (accessed on 4 August 2021).
7. Brown, M.A.; Kropf, A.J.; Paulenova, A.; Gelis, A.V. Aqueous complexation of citrate with neodymium(III) and americium(III): A study by potentiometry, absorption spectrophotometry, microcalorimetry, and XAFS. *Dalton Trans.* **2014**, *43*, 6446–6454. [CrossRef] [PubMed]
8. Moldoveanu, G.; Papangelakis, V. Recovery of rare earth elements adsorbed on clay minerals: II. Leaching with ammonium sulfate. *Hydrometallurgy* **2013**, *131–132*, 158–166. [CrossRef]
9. Moldoveanu, G.A.; Papangelakis, V.G. An overview of rare-earth recovery by ion-exchange leaching from ion-adsorption clays of various origins. *Mineral. Mag.* **2016**, *80*, 63–76. [CrossRef]
10. Shan, X.-q.; Lian, J.; Wen, B. Effect of organic acids on adsorption and desorption of rare earth elements. *Chemosphere* **2002**, *47*, 701–710. [CrossRef]
11. Evangelou, M.W.H.; Ebel, M.; Schaeffer, A. Chelate assisted phytoextraction of heavy metals from soil. Effect, mechanism, toxicity, and fate of chelating agents. *Chemosphere* **2007**, *68*, 989–1003. [CrossRef]
12. Wang, L.; Liao, C.; Yang, Y.; Xu, H.; Xiao, Y.; Yan, C. Effects of organic acids on the leaching process of ion-adsorption type rare earth ore. *J. Rare Earths* **2017**, *35*, 1233–1238. [CrossRef]
13. Huang, Y.; Wang, D.; Liu, H.; Fan, G.; Peng, W.; Cao, Y. Selective complexation leaching of copper from copper smelting slag with the alkaline glycine solution: An effective recovery method of copper from secondary resource. *Sep. Purif. Technol.* **2023**, *326*, 124619. [CrossRef]
14. Yang, J.; Montross, S.; Britton, J.; Stuckman, M.; Lopano, C.; Verba, C. Microanalytical Approaches to Characterizing REE in Appalachian Basin Underclays. *Minerals* **2020**, *10*, 546. [CrossRef]
15. Montross, S.N.; Yang, J.; Britton, J.; McKoy, M.; Verba, C. Leaching of Rare Earth Elements from Central Appalachian Coal Seam Underclays. *Minerals* **2020**, *10*, 577. [CrossRef]
16. Yang, J.; Bauer, S.; Verba, C. *Strategies to Recover Easily-Extractable Rare Earth Elements and Other Critical Metals from Coal Waste Streams and Adjacent Rock Strata Using Citric Acid*; National Energy Technology Laboratory (NETL): Pittsburgh, PA, USA; Morgantown, WV, USA, 2022.
17. Zhang, W.; Honaker, R. Characterization and recovery of rare earth elements and other critical metals (Co, Cr, Li, Mn, Sr, and V) from the calcination products of a coal refuse sample. *Fuel* **2020**, *267*, 117236. [CrossRef]
18. Bauer, S.; Yang, J.; Stuckman, M.; Verba, C. Rare Earth Element (REE) and Critical Mineral Fractions of Central Appalachian Coal-Related Strata Determined by 7-Step Sequential Extraction. *Minerals* **2022**, *12*, 1350. [CrossRef]
19. Ait Brahim, J.; Ait Hak, S.; Achiou, B.; Boulif, R.; Beniazza, R.; Benhida, R. Kinetics and mechanisms of leaching of rare earth elements from secondary resources. *Miner. Eng.* **2022**, *177*, 107351. [CrossRef]
20. Xiao, Y.; Liu, X.; Feng, Z.; Huang, X.; Huang, L.; Chen, Y.; Wu, W. Role of minerals properties on leaching process of weathered crust elution-deposited rare earth ore. *J. Rare Earths* **2015**, *33*, 545–552. [CrossRef]
21. He, Z.; Zhang, Z.; Yu, J.; Zhou, F.; Xu, Y.; Xu, Z.; Chen, Z.; Chi, R. Kinetics of column leaching of rare earth and aluminum from weathered crust elution-deposited rare earth ore with ammonium salt solutions. *Hydrometallurgy* **2016**, *163*, 33–39. [CrossRef]
22. Safari, V.; Arzpeyma, G.; Rashchi, F.; Mostoufi, N. A shrinking particle—Shrinking core model for leaching of a zinc ore containing silica. *Int. J. Miner. Process.* **2009**, *93*, 79–83. [CrossRef]
23. Faraji, F.; Alizadeh, A.; Rashchi, F.; Mostoufi, N. Kinetics of leaching: A review. *Rev. Chem. Eng.* **2020**, *38*, 113–148. [CrossRef]
24. Jander, W. Reaktionen im festen Zustande bei höheren Temperaturen. Reaktionsgeschwindigkeiten endotherm verlaufender Umsetzungen. *Z. Für Anorg. Und Allg. Chem.* **1927**, *163*, 1–30. [CrossRef]
25. Ginstling, A.; Brounshtein, B. Concerning the diffusion kinetics of reactions in spherical particles. *J. Appl. Chem. USSR* **1950**, *23*, 1327–1338.
26. Pang, Y.; Li, Q. A review on kinetic models and corresponding analysis methods for hydrogen storage materials. *Int. J. Hydrog. Energy* **2016**, *41*, 18072–18087. [CrossRef]
27. Crank, J. *The Mathematics of Diffusion*, 2nd ed.; Oxford University Press: Oxford, UK, 1975.
28. Perez, E.E.; Carelli, A.A.; Crapiste, G.H. Temperature-dependent diffusion coefficient of oil from different sunflower seeds during extraction with hexane. *J. Food Eng.* **2011**, *105*, 180–185. [CrossRef]

29. Viegas, R.; Campinas, M.; Costa, H.; Rosa, M. How do the HSDM and Boyd's model compare for estimating intraparticle diffusion coefficients in adsorption processes. *Adsorption* **2014**, *20*, 737–746. [CrossRef]
30. Plazinski, W.; Rudzinski, W. Kinetics of Adsorption at Solid/Solution Interfaces Controlled by Intraparticle Diffusion: A Theoretical Analysis. *J. Phys. Chem. C* **2009**, *113*, 12495–12501. [CrossRef]
31. Boyd, G.E.; Schubert, J.; Adamson, A.W. The Exchange Adsorption of Ions from Aqueous Solutions by Organic Zeolites. I Ion-exchange Equilibria1. *J. Am. Chem. Soc.* **1947**, *69*, 2818–2829. [CrossRef]
32. Reichenberg, D. Properties of Ion-Exchange Resins in Relation to their Structure. III. Kinetics of Exchange. *J. Am. Chem. Soc.* **1953**, *75*, 589–597. [CrossRef]
33. Turner, B.D.; Henley, B.J.; Sleap, S.B.; Sloan, S.W. Kinetic model selection and the Hill model in geochemistry. *Int. J. Environ. Sci. Technol.* **2015**, *12*, 2545–2558. [CrossRef]
34. Dada, A.O.; Adekola, F.A.; Odebunmi, E.O. A novel zerovalent manganese for removal of copper ions: Synthesis, characterization and adsorption studies. *Appl. Water Sci.* **2017**, *7*, 1409–1427. [CrossRef]
35. Tsibranska, I.; Hristova, E. Comparison of different kinetic models for adsorption of heavy metals onto activated carbon from apricot stones. *Bulg. Chem. Commun.* **2011**, *43*, 370–377.
36. Silva, A.M.N.; Kong, X.; Hider, R.C. Determination of the pKa value of the hydroxyl group in the α-hydroxycarboxylates citrate, malate and lactate by 13C NMR: Implications for metal coordination in biological systems. *BioMetals* **2009**, *22*, 771–778. [CrossRef] [PubMed]
37. Shi, Q.; Zhao, Y.; Meng, X.; Shen, L.; Qiu, G.; Zhang, X.; Yu, H.; He, X.; He, H.; Zhao, H. Column leaching of ion adsorption rare earth ore at low ammonium concentration. *J. Mater. Res. Technol.* **2022**, *19*, 2135–2145. [CrossRef]
38. Goyne, K.W.; Brantley, S.L.; Chorover, J. Rare earth element release from phosphate minerals in the presence of organic acids. *Chem. Geol.* **2010**, *278*, 1–14. [CrossRef]
39. Bolan, N.S.; Naidu, R.; Mahimairaja, S.; Baskaran, S. Influence of low-molecular-weight organic acids on the solubilization of phosphates. *Biol. Fertil. Soils* **1994**, *18*, 311–319. [CrossRef]
40. Uddin, F. Clays, nanoclays, and montmorillonite minerals. *Metall. Mater. Trans. A* **2008**, *39*, 2804–2814. [CrossRef]
41. Kirby, C.S.; Rimstidt, J.D. Interaction of municipal solid waste ash with water. *Environ. Sci. Technol.* **1994**, *28*, 443–451. [CrossRef]
42. Dijkstra, J.J.; van der Sloot, H.A.; Comans, R.N.J. The leaching of major and trace elements from MSWI bottom ash as a function of pH and time. *Appl. Geochem.* **2006**, *21*, 335–351. [CrossRef]
43. Brady, J.; Cherniak, D. Diffusion in Minerals: An Overview of Published Experimental Diffusion Data. *Rev. Mineral. Geochem.* **2010**, *72*, 899–920. [CrossRef]

separations

MDPI

Article

Enhancing Trace Metal Extraction from Wastewater: Magnetic Activated Carbon as a High-Performance Sorbent for Inductively Coupled Plasma Optical Emission Spectrometry Analysis

Sergio J. Abellán-Martín [1], David Villalgordo-Hernández [2], Miguel Ángel Aguirre [1,*], Enrique V. Ramos-Fernández [2], Javier Narciso [2,3] and Antonio Canals [1,*]

1 Department of Analytical Chemistry and Food Science, University Institute of Materials, Faculty of Science, University of Alicante, P.O. Box 99, 03080 Alicante, Spain; sergio.abellan@ua.es

2 Department of Inorganic Chemistry, University Institute of Materials, Faculty of Science, University of Alicante, P.O. Box 99, 03080 Alicante, Spain; david.villalgordo@ua.es (D.V.-H.); enrique.ramos@ua.es (E.V.R.-F.); narciso@ua.es (J.N.)

3 Instituto de Investigación Sanitaria y Biomédica de Alicante (ISABIAL), 03010 Alicante, Spain

* Correspondence: aguirre.pastor@ua.es (M.Á.A.); a.canals@ua.es (A.C.)

Abstract: A new fast, sensitive, and environmentally friendly analytical method has been developed for the simultaneous determination of Ba, Be, Cd, Cr, Cu, Fe, Hg, Mn, Ni, Pb, and Zn in wastewater samples using inductively coupled plasma optical emission spectroscopy (ICP OES). A preconcentration step using a magnetic dispersive solid-phase extraction (MDSPE) technique with a new magnetic sorbent was performed. The new sorbent material was a carbon containing magnetic cobalt and nitrogen groups. This material was synthetized using controlled pyrolysis of a zeolitic imidazolate framework (i.e., ZIF-67). In order to optimize the experimental parameters that affect the MDSPE procedure, a multivariate optimization strategy, using Plackett–Burman and circumscribed central composite designs (CCD), was used. The method has been evaluated employing optimized experimental conditions (i.e., sample weight, 10 g; sample pH, 7.6; amount of sorbent, 10 mg; dispersive agent, vortex; complexing agent concentration, 0.5%; ionic concentration, 0%; eluent, HCl; eluent concentration, 0.5 M; eluent volume, 300 µL; elution time, 3 min and extraction time, 3 min) using external calibration. Limits of detection (LODs) in a range from 0.073 to 1.3 µg L^{-1} were obtained, and the repeatability was evaluated at two different levels, resulting in relative standard deviations below 8% for both levels (n = 5). An increase in the sensitivity was observed due to the high enrichment factors (i.e., 3.2 to 13) obtained compared with direct ICP OES analysis. The method was also validated through carrying out recovery studies that employed a real wastewater sample and through the analysis of a certified reference material (ERM®-CA713). The recovery values obtained with the real wastewater were between 94 and 108% and between 90 and 109% for the analysis of ERM®-CA713, showing negligible matrix effects.

Keywords: magnetic dispersive solid-phase extraction (MDSPE); magnetic activated carbon; trace metal analysis; multivariate optimization; wastewater samples; inductively coupled plasma optical emission spectrometry (ICP OES)

1. Introduction

An important effect of industrial processes and human activities is the pollution of several ecosystems [1–3]. In this regard, heavy metals pollution has risen in recent years and has become an ongoing issue for environmental organizations, being one of the most serious ecological concerns at present [4,5]. These pollutants may enter the human body through air, water, food, or absorption through the skin [6]. It is well known that exposure to certain heavy metals has been linked to a number of disorders that are harmful to human health since, some of them are not metabolized and can accumulate in soft tissues [7,8]. According to the World Health Organization (WHO), long-term exposure to environmental pollution, such as air, soil, and water pollution, causes around 25% of the illnesses that humans face today [9]. Therefore, the determination of these toxic metal ions in water and aquatic ecosystems is of great importance, as these contaminants are detrimental to the environment and have serious impacts on human health [10].

Various analytical techniques such as inductively coupled plasma optical emission spectrometry (ICP OES) [11–13], inductively coupled plasma mass spectrometry (ICP-MS) [14–16], electrothermal atomic absorption spectrometry (ETAAS) [17–19], flame atomic absorption spectrometry (FAAS) [20–22], and atomic fluorescence spectrometry (AFS) [23,24] have been applied for the determination of trace amounts of heavy metal ions in water and biological samples. Because of its superior sensitivity, broad dynamic linear range, reduced detection limits, excellent accuracy, and simultaneous multi-elemental analysis capacity, ICP OES is one of the most popular analytical techniques for elemental analysis [11,25]. Despite these benefits, trace metals concentration in real samples is extremely low and the complexity of matrices in real samples may substantially interfere with the determination of analytes [26,27]. Therefore, to solve such problems, it is necessary to develop simple, selective, reliable, efficient, and eco-friendly sample pretreatment methods for the extraction of trace amounts of heavy metals in environmental and biological samples before ICP OES determination [28]. Under these conditions, a separation and enrichment step such as liquid–liquid extraction (LLE) [29–31], coprecipitation [32–34], filtration [35], ion exchange [36–38], cloud point extraction (CPE) [39,40], or solid-phase extraction (SPE) [41–44] have been used prior to its determination.

Among the above-mentioned separation and preconcentration techniques, solid-phase extraction (SPE) is one of the most widely used sample pretreatment techniques [45]. SPE has some advantages over other sample pretreatment techniques due to its simplicity, reduced extraction time, greater enrichment factors, cheaper cost, and the reduction in the consumption of organic solvents [46,47]. The fundamental concept of SPE is the transfer of analytes from the liquid phase to the solid phase's adsorption sites where they remain during the extraction process [48]. Then, both phases are separated by centrifugation or simply recovered by flotation [49,50]. After adsorption, the analytes retained in the solid sorbent can be eluted [51], thermally desorbed [52], or directly determined using a suitable analytical technique [53]. SPE has been extensively employed for the preconcentration of metals in polluted waters [54].

Despite all these advantages, there is growing interest in developing ecological analytical methods based on the Green Analytical Chemistry principles [55]. The Green Analytical Chemistry principles and the processes which are adhered to them aim to replace harmful chemicals, reduce waste in laboratories, and, as a result, miniaturize traditional analytical procedures. Furthermore, in recent years, interest in the sample preparation step has increased in relation to the principles of Green Analytical Chemistry. For that reason, the continuous search for innovative sample preparation methods has led to the creation of new procedures such as solid-phase microextraction (SPME) [56] and dispersive solid-phase extraction (DSPE) [57]. In DSPE, the solid sorbent is dispersed in the solution. This strategy encourages quick interaction of the analytes with the sorbent, minimizing the consumption of sorbent and leading to a faster extraction process in comparison with traditional SPE [58]. New advances in sample preparation that employ DSPE try to reduce the consumption of sorbents by using smaller amounts (i.e., microgram); this tendency has led to the development of dispersive micro-solid-phase extraction (DμSPE) [59].

Magnetic materials have recently been reported to be used as sorbents in several papers [6,60,61]. Magnetic dispersive solid-phase extraction (MDSPE) is gaining popularity because it decreases sample preparation time and allows for easier sorbent manipulation [62,63]. In MDSPE, the magnetic sorbent material is dispersed into the liquid phase; after the extraction, it is simply removed from the sample solution using a magnet [64], avoiding time-consuming filtration or centrifugation steps to carry out the phase separation [61,65,66]. The sorbents decorated with magnetic solids simultaneously combine the exceptional sorbent capability and the easy sorbent handling, reducing analysis time, sorbent amounts, and the consumption of solvents [67,68]. Finally, analytes can be desorbed employing an eluent solvent or heating for further analysis [69].

One of the most important MDSPE-related research areas at present is the quest for novel magnetic sorbent materials with high selectivity and strong adsorption capacity that are chemically and thermally stable, inexpensive, and environmentally benign [14,70].

Different magnetic sorbents have been used in magnetic dispersive solid-phase extraction, such as inorganic oxides, porous polymers, molecularly imprinted polymers (MIP), biosorbents, or carbon nanomaterials [71,72]. In addition, magnetic composites based on zeolitic materials have been described for the extraction and preconcentration of organic compounds and metals from different samples, including water, environment, food, and biological samples [6,73–75].

The aim of this work is to develop a new analytical method for the determination of trace metals (i.e., Ba, Be, Cd, Cr, Cu, Fe, Hg, Mn, Ni, Pb, and Zn) in wastewater samples through MDSPE using a new magnetic material developed specifically for this application. This material is carbon, containing magnetic cobalt and nitrogen groups. These groups can serve as anchoring centers for the metals and have been prepared using controlled pyrolysis of a zeolitic imidazolate framework (ZIF). These materials are composed of tetrahedrally coordinated transition metal ions (e.g., Fe, Co, Cu, Zn) that are connected by imidazolate linkers; this is further carbonized in order to obtain Co-doped magnetic carbon. More information about the employed material can be found in the bibliography [76]. To the best of our knowledge, the new magnetic material has not been employed for the determination of trace metals before.

2. Materials and Methods

2.1. Instrumentation

All the measurements were carried out using an Agilent 720 inductively coupled plasma optical emission spectrometer (Agilent Technologies, Melbourne, Australia). The operating conditions employed using the ICP OES are shown in Table S1. The instrumental parameters were optimized, accomplishing the maximum signal intensity for the set of emission lines evaluated. In addition, the sample uptake rate (i.e., 100 $\mu L\ min^{-1}$) was the minimum liquid flow required in order to measure all emission lines analyzed. A pH meter Basic 20 (Crison Instrument, Barcelona, Spain) provided with a combined glass electrode was used for the pH measurements. In order to disperse the extractant into the sample, a Vortex Reax Top (Heidolph Instruments, Schwabach, Germany) was used. A neodymium magnet (1.32–1.37 T) was used to bring about the phase separation (Supermagnete, Gottmadingen, Germany). NemrodW statistical software (NemrodW® v.2007/2010, LPRAI, Marseille, France) and Statgraphics® Centurion (Statpoint Technologies, Warreton, VA, USA) were employed to construct the experimental designs and evaluate the optimization results and the desirability function. NemrodW was used to evaluate the screening and the central circumscribed design of the developed method and Statgraphics® Centurion was employed to construct the desirability function to obtain a common experimental condition for all the analytes evaluated.

Characterization of the porosity of the samples was performed using N_2 adsorption–desorption isotherms as well as CO_2 adsorption isotherms [77]. Before measurement, samples were outgassed at 150 °C during 4 h (Autosorb degasser, Quantachrome, Boynton Beach, FL, USA). The nitrogen adsorption–desorption isotherms were measured at −196 °C (Autosorb-6, Quantachrome). The Brunauer–Emmett–Teller (BET) equation was utilized for the estimation of the surface area, which was determined from the N_2 adsorption branch. The number of points used to apply the BET equation was higher than five, and the BET constant was always positive. The Dubinin–Radushkevich method was utilized to obtain the micropore volume (V_{micro}) and determine whether any diffusional restrictions in the adsorption took place. Pore volume has been obtained at $P/P_0 = 0.95$. All data treatment related to sorbent characterization was carried out using the software Quadrawin (Quantachrome). The same device was used for the CO_2 adsorption isotherms at 0 °C. No diffusional restrictions were observed. Powder X-ray diffraction was utilized for the identification of crystallographic phases using a D8-Advanced diffractometer with a Goëbel mirror and a Kristalloflex K 760-80F X-ray generation system, fitted with a Cu cathode and a Ni filter (Bruker, Billerica, MA, USA). Spectra were registered between 5° and 80° with an angular step of 0.05° and a three second step time. X-ray photoelectron spectroscopy (XPS) analysis was carried out in a VG-Microtech Multilab device with a MgKalpha (Hv:1253.6 eV) radiation source, pass energy of 50 eV, and background pressure of $5 \cdot 10^{-7}$ Pa (VG-Microtech, Uckfield, UK). The spectra were carefully deconvoluted and the peak areas estimated through integration of each peak, prior subtraction of a Shirley background, and fitting the experimental peaks to a 30/70 proportion of a combination of Lorentzian/Gaussian lines. Field emission scanning electron microscopy (FESEM model Merlin VP Compact, Zeiss, Jena, Germany) with energy-dispersive X-ray spectroscopy (EDS, model Quantax 400, Bruker) in both backscattered electron (BSE) and secondary electron (SE) modes were employed to study the morphology of the samples. Raman spectroscopy characterization was carried out in an NRS-5100 device from Jasco (Tokyo, Japan) using a 633 nm He laser. Magnetic susceptibility measurements were carried out at 300 K in a magnetic field ranging from −30 to 30 kOe using a MPMS XL (SQUID) magnetometer from Quantum Design (San Diego, CA, USA). Finally, the zeta potential was measured using a Colloid Metrix Stabino II instrument (Meerbusch, Germany).

2.2. Samples and Reagents

Standard reference solutions of Hg and Be used for ICP OES calibration and for addition and recovery experiments were purchased from High Purity Standard (Charleston, SC, USA) with a concentration of 1000 mg L^{-1}. Standard reference solutions of Ba, Cd, Cr, Cu, Fe, Mn, Ni, Pb, and Zn were obtained from Merck IV solution of 1000 mg L^{-1} (Merck, Darmstadt, Germany). Working solutions of analytes were prepared using appropriate dilution of the commercial standard reference solutions. To prepare the buffer solutions during the optimization and measurement process, H_3PO_4 85% purity (Scharlau, Barcelona, Spain) and KH_2PO_4 and K_2HPO_4 pro-analysis (Merck) were employed. Sodium diethyldithiocarbamate (DDTC, 99%, Merck) was used as a complexing agent. HCl 30% purity (Merck) and HNO_3 65% purity (Scharlau) were employed as eluent phases during the measurement and optimization process. During the optimization process, NaCl pro-analysis (Scharlau) was used to adjust the ionic strength of the samples. All the experiments were prepared using ultrapure water with resistivity higher than 18.0 MΩ cm obtained using a PURELAB flex 3 purification system (Elga LabWater, High Wycombe, UK). Reagents employed during the synthesis of ZIF-67 were provided by Merck in analytical grade and were used without further purification.

The standard reference material employed in this work was ERM®-CA713 wastewater (Joint Research Center, European Commission, Geel, Belgium). A real water sample used was obtained from a wastewater treatment plant (Aguas Municipalizadas de Alicante E.M., Spain). Sample was collected in amber glass containers and then stored in the dark at 4 °C. To perform the analysis, the sample supernatant was collected and measured (i.e., dissolved metals) after decantation during 24 h without any sample digestion or filtration step.

2.3. Synthesis of the Magnetic Sorbent

The synthetic procedure followed the classical ZIF-67 synthesis [78,79], in which 144 mmol of 2-Methylimidazole were dissolved in 50 mL of deionized water. This solution was blended with a 25 mL deionized water solution in which 12 mmol of cobalt acetate were previously dissolved. The mixture was stirred vigorously for 5 min and kept at room temperature for 72 h. ZIF-67 was recovered by centrifugation at 12,000 rpm and washed three times with methanol. Then, the powder was dried at 60 °C for 24 h in a conventional oven. Lastly, the ZIF was put in a tubular furnace under a 100 mL min^{-1} N_2 dynamic atmosphere, heated up at 5 °C min^{-1}, and carbonized at 900 °C for 2 h. The carbon material was then washed with HCl following the procedure of Villalgordo-Hernández et al. [76] (i.e., ZIF67C_900_l meaning activated carbon obtained by carbonization at 900 °C and washed once with HCl). Previously, a screening study of different magnetic carbons was conducted, finding that this material was the most suitable for this specific application.

2.4. MDSPE Procedure

For the magnetic microextraction process, the first step was to adjust the pH of the sample or standards at 7.6 using the necessary amount of KH_2PO_4 and K_2HPO_4. Subsequently, DDTC was added as a complexing agent until its concentration was 0.5%. Once the sample (or standard) was prepared, it was transferred (i.e., 10 g) to a glass centrifugation tube and 10 mg of the magnetic solid extractant (ZIF67C_900_l) was introduced in the tube that contains the sample or standard. Then, the mixture was shaken using the vortex for 3 min to extract the analytes. After shaking, the supernatant (i.e., sample) was removed and the sorbent was washed 3 times with 1 mL of ultrapure water and the wash water was removed. Afterwards, 300 µL of HCl 0.5 M was added to the sorbent in order to elute the analytes. The elution process was carried out using the vortex for 3 min. Finally, the eluate was removed from the solution containing the sorbent and measured directly with the ICP OES. During the extraction, washing, and elution process, the neodymium magnet was used to ease the separation step between the solid and liquid phases. The entire MDSPE process is shown in Figure 1.

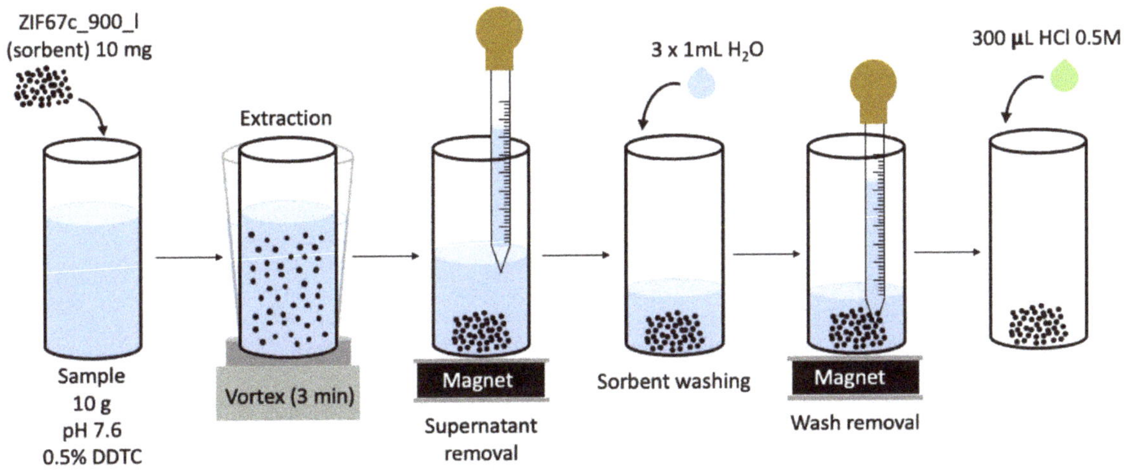

Figure 1. *Cont.*

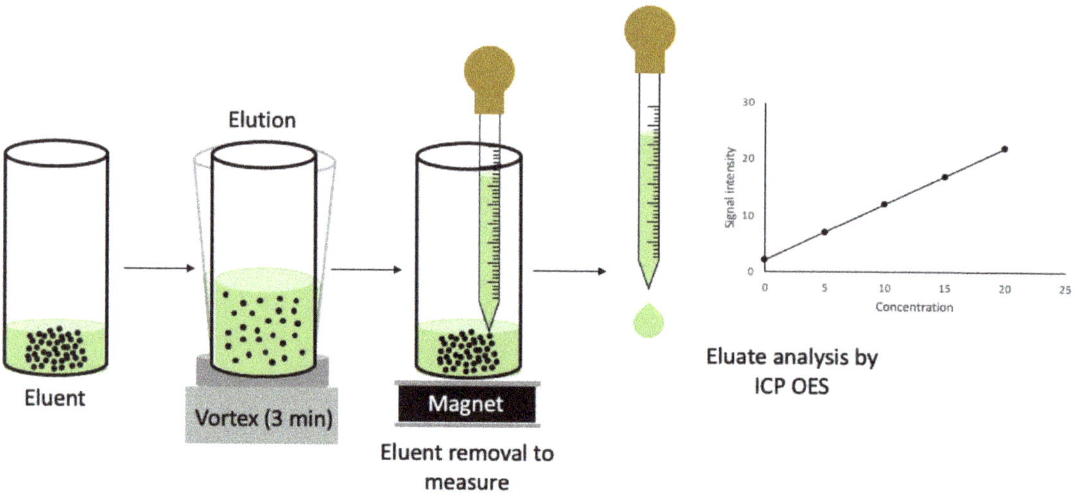

Figure 1. Schematic representation of MDSPE for preconcentration of trace metals using the optimized experimental conditions (i.e., (i) sample weight, 10 g; (ii) sample pH, 7.6; (iii) amount of sorbent, 10 mg; (iv) dispersion mode, vortex; (v) complexing agent concentration, 0.5%; (vi) ionic concentration, 0%; (vii) eluent, HCl; (viii) eluent concentration, 0.5 M; (ix) eluent volume, 300 μL; (x) extraction time, 3 min; (xi) elution time, 3 min).

3. Result and Discussion

3.1. Characterization of ZIF67C_900_1

As presented in Figure S1, three diffraction peaks appear at 44.2°, 51.5°, and 75.9° in the XRD pattern, which could be indexed to Co(0) (JCPDS card No. 15-0806). Moreover, the peak found at 25.85° could be assigned to the (002) plane of graphite (JCPDS card No. 89-8487). In all cases, well-defined peaks can be observed, which indicate that both carbon and cobalt are crystalline, but there are crystallite sizes of less than 100 nm, since the peaks are very broad.

The N_2 adsorption isotherm (Figure S2) of the original carbon and the washed one are very similar; in both cases, they show a very wide distribution of porosity, which is easily deduced from the wide shoulder that shows the adsorption. Although the specific surface area is not very high in either case (400 and 480 $m^2\ g^{-1}$), the washed sample has an area that is 20% higher. It should also be noted that it presents a certain hysteresis (desorption branch above the adsorption branch); this indicates that we have ink-bottle-type pores that have been generated by acid washing the carbon due to the extraction of the unbound cobalt particles [76].

The secondary electron micrograph (Figure 2) shows the cobalt-doped carbon crystallites. The microstructure reminds us of the original ZIF, where there has simply been a contraction of the particles without losing their original cubo-octahedral shape [80]. In addition, it is possible to observe from the secondary electron micrograph that the particle size of the ZIF is 1 μm.

The XPS spectra shown in Figure S3 are very revealing. Firstly, we will analyze Co. The XPS spectrum shows the 2p1/2 transition where a peak associated with Co(0) and a peak at higher energy associated with Co(II) can be clearly seen. The presence of the satellites is unequivocal that we have Co(II) and not Co(III). A fact to highlight is that the presence of Co(0), which is what we have seen by other techniques, indicates that we only have a small nanometric layer of cobalt oxide. Regarding the nitrogen spectrum, we observe the 1s transition, where three peaks can be seen that we can associate with three N positions in the graphene sheet, as indicated in the figure.

Figure 2. Secondary electron micrograph of ZIF67C_900_1.

The Raman spectrum shows four easily identifiable bands [81] as shown in Figure S4. First of all, highlight the two bands called G and 2D, which indicate that the carbon is not turbostratic but rather graphitic, as the intensity of the G band is much higher than that of the carbon. The 2D band indicates that there are stacks of more than 10 layers. The presence of the D band, also called the defects, is indicative that the graphene sheets have a considerable number of defects, and that there are areas where the carbon is not perfectly ordered.

Figure S5 shows the magnetic susceptibility of the studied ZIF67C_900_1 material at 300 K in a magnetic field that ranges from −30 to 30 kOe. At the magnetization curve it is possible to observe a sigmoid shape that passes approximately through the origin having an extremely narrow hysteresis loop. In addition, the coercivity field (H_C) and residual magnetization (M_R) values are significantly small while saturation magnetization (Ms) is relatively high, the values are 200 KOe, 3.91 emu g^{-1}, and 15.34 emu g^{-1}, respectively, showing a ferromagnetic behavior [82].

To investigate the surface charges of the ZIF67C_900_1 in the same matrix sample performed the MDSPE procedure, their zeta potentials were measured at different sample pH. The zeta potential of ZIF67C_900_1 as function of the sample pH of the suspension is given in Figure S6. In the suspension with an initial pH of 4.0 the zeta potential was 20.3 mV. As can be seen it continuously decreases when the pH is increased up to a pH value of 7.0, where the zeta potential is stabilized; the isoelectric point of the magnetic material is 5.0. Therefore, the surface of the magnetic material is positively charged at pH values below the isoelectric point and negatively charged above this point. The negative charge surface of the magnetic sorbent is of great importance, since it affects the extraction process of positively charged metals. For that reason, the optimum pH value was 7.6 (Section 3.2) where the charge was negative (i.e., the material was deprotonated), increasing the interactions with metals and, thus, increasing the extraction efficiency.

3.2. MDSPE Multivariate Optimization

Since MDSPE is affected by many experimental variables, it is necessary to optimize the entire extraction process. Many experiments would have been necessary to optimize all experimental variables if they had been done through univariate optimization, increasing the consumption of sample, reagents, energy and time. To overcome this fact, multivariate optimization has recently emerged as an alternative to univariate optimization, since the number of experiments was significantly reduced. For that reason, a multivariate optimization design has been employed in order to obtain the optimal experimental conditions, considering the simultaneous determination of all analytes. The optimization process was carried out using a Plackett–Burman design as a screening approach to distinguish between significant and non-significant factors, followed by a central composite design (CCD) in order to obtain the optimal values of the significant factors. To perform the MDSPE optimization experiments a standard of 500 $\mu g\ L^{-1}$ containing the elements was used. Then, the analytical signals obtained measuring with ICP OES were individually used as response function for optimization. In this way, individual optimum conditions were obtained for each analyte. In order to obtain a global optimum condition for all the analytes, a desirability function was employed to extract all the analytes with the best conditions.

3.2.1. Screening Study

The eleven factors that affect the MDSPE were selected based on previous experience and experiments carried out by our research group. The factors evaluated in the Plackett–Burman design were the following: (i) sample weight, (ii) sample pH, (iii) amount of sorbent, (iv) dispersion mode, (v) complexing agent concentration, (vi) ionic concentration, (vii) eluent, (viii) eluent concentration, (ix) eluent volume, (x) extraction time, and (xi) elution time. The positive and negative levels of each factor are shown in Table S2. A matrix of twelve experiments was constructed and randomly performed in order to obtain the analytical signals.

Once the obtained signals were used in the Plackett–Burman design, Pareto charts were constructed, showing the significant and non-significant factors (Figure S7). The length of each bar in these Pareto charts is related to the relative importance of that factor in the MDSPE approach. Furthermore, the bars to the right show a positive effect in the response, when the value of the factor is increased from the lower to the upper level. On the other hand, the bars to the left indicate a negative effect. Additionally, the bars that surpass reference vertical lines (dashed lines) are 95% likely to be significant [83].

Regarding the results in Figure S7, it can be seen that only the eluent (i.e., HCl), the dispersion mode (i.e., vortex), the sample pH, and the eluent volume significantly affect the metal extraction procedure for the majority of the analytes that were evaluated. In addition, MDSPE is favored with the positive effects of the sample weight (10 g), the amount of sorbent (10 mg), and the extraction and elution time (3 min). On the other hand, it is favored with the negative effect of the complexing agent concentration (0.5%), the ionic concentration (0%), and the eluent concentration (0.5 M). Among the four factors that significantly affect the MDSPE, two of them are binary (i.e., dispersion mode and eluent) and the others are continuous (i.e., sample pH and eluent volume). Therefore, the binary and the non-significant factors have been directly fixed at the most favorable levels for the central composite design.

3.2.2. Central Composite Design (CCD)

Once the significant parameters affecting the MDSPE were determined, a central composite design (CCD) was formulated to obtain the optimum sample pH and eluent volume for each element. In this design, there were five levels (Table S3). The response surfaces obtained are shown in Figure S8 and the optimum conditions for each element are shown in Table S4.

3.2.3. Desirability Function

Due to the different results obtained in the sample for the optimum pH and eluent volume for each element, it is necessary to find the same optimum conditions for all the analytes evaluated. For that reason, the desirability function was employed to obtain a common set of optimum extraction conditions to facilitate the experimental procedure. Figure 3 shows the optimum values of sample pH and eluent volume for all the elements by applying the desirability function. In the graphics obtained using this function, the closest result to a value of 1 is the best fit for all the elements. Therefore, once the desirability function was applied, the obtained optimized conditions for simultaneous extraction of all evaluated metals were the ones shown in Table S5. It was possible to observe that the optimum pH value obtained (i.e., 7.6) was above the isoelectric point (i.e., pH = 5), as predicted in Section 3.1. This was due to the fact that, above the isoelectric point, the material was negatively charged; this affects the extraction process by increasing the affinity for metals (i.e., positively charged).

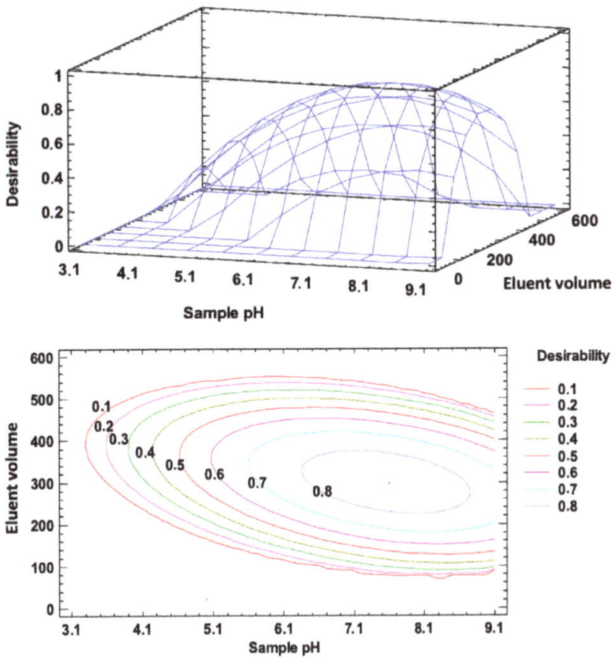

Figure 3. Graphics with obtained optimum results of sample pH and eluent volume applying the desirability function. Experimental fixed conditions: (i) sample weight, 10 g; (ii) amount of sorbent, 10 mg; (iii) dispersion mode, vortex; (iv) complexing agent concentration, 0.5%; (v) ionic concentration, 0%; (vi) eluent, HCl; (vii) eluent concentration, 0.5 M; (viii) extraction time, 3 min; (ix) elution time, 3 min.

3.3. Sorbent Reutilization Study

The possibility of reusing the sorbent was also studied. To that end, five consecutive extractions were carried out in triplicate, employing the same MDSPE procedure; then, the magnetic sorbent was washed three times with 1 mL of ultrapure water. Finally, the magnetic composite was dried at 130 °C for 1 h after each extraction process in order to eliminate the excess water. The experiments were performed using a standard of 5 μg L^{-1}. As shown in Figure 4, the signal intensity decreases after each extraction, which may be due to the corrosive nature of the acid used in the elution process and the washing out of the functional groups. In addition, the phase separation using the magnet took longer time after each extraction. For these reasons, the material cannot be used more than once. Nonetheless, the synthesis of ZIF67C_900_1 is very simple and inexpensive. Thus, the disadvantage of not being able to reuse the material is not a limiting factor in carrying out the developed method. In addition, new synthesis routes are currently being investigated in our laboratory to provide a reusable material with the same extraction efficiency.

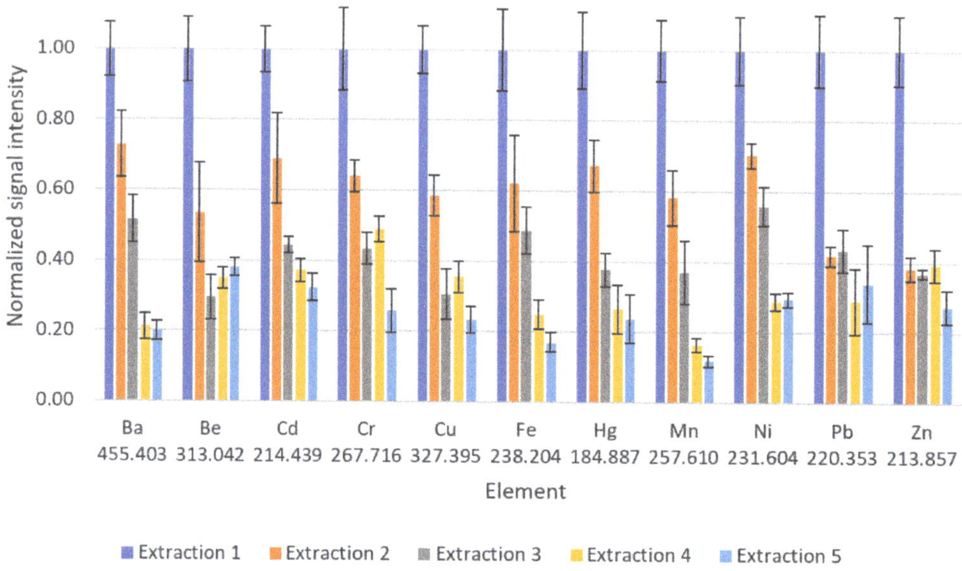

Figure 4. Study of sorbent reusability using the same sorbent in five consecutive extractions of 5 μg L^{-1} standard solution. The sorbent was washed three times with 1 mL of ultrapure water and heated at 130° for 1 h; it was dried between each extraction. The error bars were evaluated through the standard deviation of three replicates. Experimental conditions used: (i) sample weight, 10 g; (ii) sample pH, 7.6; (iii) amount of sorbent, 10 mg; (iv) dispersion mode, vortex; (v) complexing agent concentration, 0.5%; (vi) ionic concentration, 0%; (vii) eluent, HCl; (viii) eluent concentration, 0.5 M; (ix) eluent volume, 300 μL; (x) extraction time, 3 min; (xi) elution time, 3 min.

3.4. Method Validation

All the analytical figures of merit of the proposed method (Table 1) were assessed under optimum conditions, as shown in Table S5. For the purposes of comparison, the proposed method (i.e., MDSPE-ICP OES) was contrasted with the direct ICP OES analysis (i.e., without the MDSPE procedure). In the case of the proposed method, the calibration curves showed coefficients of determination (R^2) that ranged from 0.993 to 0.9996 (i.e., $N = 7$), demonstrating good linearity. In the direct ICP OES analysis, good linearities were also obtained (i.e., 0.98–0.9994). Five independent measurements of samples spiked at 25 and 300 μg L^{-1} of each target element were used to assess the repeatability of the proposed method. The obtained relative standard deviations were between 4 and 8%. The sensitivity was also evaluated for both methods and compared through the enrichment factor. Sensitivities ranged between 0.39 and 877 cps μg^{-1} L in the direct ICP OES analysis. A significant increase in the sensitivity was observed in the proposed method, obtaining values that ranged from 3.03 to 7103 cps μg^{-1} L. Finally, in order to compare the sensitivity values, the enrichment factor was calculated using the ratio between the obtained sensitivities with and without the MDSPE procedure; the results of enrichment factors were between 3.2 and 13. These values of the enrichment factor enhanced the significant increase in the sensitivity of the applied extraction method.

Table 1. Analytical figures of merit for direct ICP OES and the MDSPE-ICP OES method, developed for the analysis of water samples.

Direct ICP OES

Emission Line (nm)	Working Range ($\mu g\ L^{-1}$)	R^2	Sensitivity [a] ($cps\ \mu g^{-1}\ L$)	Repeatability (RSD%)	LOD ($\mu g\ L^{-1}$)	LOQ ($\mu g\ L^{-1}$)	EF
Ba 455.403	250–1000	0.998	534 ± 17	-	2.6	8.8	-
Be 313.042	250–1000	0.998	877 ± 28	-	0.39	1.3	-
Cd 214.439	250–1000	0.998	8.3 ± 0.2	-	1.5	5.0	-
Cr 267.716	250–1000	0.9994	11.37 ± 0.19	-	1.3	4.4	-
Cu 327.395	250–1000	0.998	11.0 ± 0.4	-	1.2	3.9	-
Fe 238.204	250–1000	0.998	9.1 ± 0.2	-	9.8	33	-
Hg 184.887	250–1000	0.9991	2.86 ± 0.06	-	5.1	17	-
Mn 257.610	250–1000	0.996	34.3 ± 1.6	-	0.82	2.7	-
Ni 231.604	250–1000	0.98	1.24 ± 0.12	-	9.9	33	-
Pb 220.353	250–1000	0.98	0.39 ± 0.04	-	21	70	-
Zn 213.857	250–1000	0.992	25.0 ± 1.6	-	0.81	2.7	-

MDSPE-ICP OES [d]

Emission line (nm)	Working range ($\mu g\ L^{-1}$)	R^2	Sensitivity [a] ($cps\ \mu g^{-1}\ L$)	Repeatability (RSD%) [b]		LOD ($\mu g\ L^{-1}$)	LOQ ($\mu g\ L^{-1}$)	EF [c]
				25 ($\mu g\ L^{-1}$)	300 ($\mu g\ L^{-1}$)			
Ba 455.403	0.5–500	0.9991	7103 ± 93	5	5	0.085	0.29	13
Be 313.042	0.5–500	0.997	5638 ± 131	7	6	0.073	0.24	6.4
Cd 214.439	0.5–500	0.993	84 ± 3	6	4	0.12	0.41	10
Cr 267.716	0.5–500	0.998	50.2 ± 1.0	7	6	0.14	0.45	4.4
Cu 327.395	0.5–500	0.998	85.9 ± 1.5	6	6	0.10	0.34	7.8
Fe 238.204	5–500	0.997	64.6 ± 1.8	8	5	1.3	4.3	7.1
Hg 184.887	0.5–500	0.997	33.8 ± 0.8	7	6	0.13	0.43	12
Mn 257.610	0.5–500	0.997	109 ± 3	6	6	0.11	0.37	3.2
Ni 231.604	5–500	0.998	8.23 ± 0.16	7	7	1.1	3.8	6.6
Pb 220.353	5–500	0.9996	3.03 ± 0.03	8	7	1.1	3.8	7.9
Zn 213.857	1–500	0.998	92.2 ± 1.6	8	6	0.22	0.75	3.7

[a] Slope of the calibration curve ± standard deviation ($N = 7$). [b] Repeatability of the method evaluated at two levels through the RSD (n = 5). [c] Enrichment factor (EF): ratio between the slope with and without the MDSPE procedure. [d] Experimental conditions used: (i) sample weight, 10 g; (ii) sample pH, 7.6; (iii) amount of sorbent, 10 mg; (iv) dispersion mode, vortex; (v) complexing agent concentration, 0.5%; (vi) ionic concentration, 0%; (vii) eluent, HCl; (viii) eluent concentration, 0.5 M; (ix) eluent volume, 300 µL; (x) extraction time, 3 min; (xi) elution time, 3 min.

Limits of detection (LODs) and limits of quantification (LOQs) were calculated according to the guidelines established by Eurachem [84], from the value of the standard deviation of 10 measurements of the blank signal. This value is divided by the sensitivity (i.e., slope of the calibration line) and multiplied by factor k. The k value established by the IUPAC is 3 for the limit of detection and 10 for the limit of quantification. As can be seen, the LOD values were found to be between 0.073 and 0.22 $\mu g\ L^{-1}$ for all the analytes except for Fe, Ni, and Pb, with a LODs of 1.3, 1.1 and 1.1 $\mu g\ L^{-1}$, respectively. The limits of detection obtained in direct ICP OES analysis ranged from 0.39 to 21 $\mu g\ L^{-1}$. In other words, the LOD values obtained performing the MDSPE procedure were 14 times lower on average in comparison with those obtained through direct ICP OES analysis. In addition, the LOD values obtained with the developed method are on average 17 times lower than those established in the official method of the Environmental Protection Agency (EPA) (i.e., method 200.7). Finally, a certified reference material (ERM-CA713) was analyzed using the optimal conditions of the developed method in order to evaluate the method validation. The certified reference material analysis obtained recovery values between 90 and 109%; these results demonstrate the reliability of the method which was developed for the analysis of trace metals in wastewater samples. Table 2 shows the results of the certified reference material analysis.

Table 2. Results of certified reference material analysis [a].

Analyte	Cd	Cr	Cu	Fe	Hg	Mn	Ni	Pb
Certified value ($\mu g\ L^{-1}$) [b]	5.09 ± 0.20	20.9 ± 1.3	101 ± 7	445 ± 27	1.84 ± 0.11	95 ± 4	50.3 ± 1.4	49.7 ± 1.7
Found value ($\mu g\ L^{-1}$) [c]	4.8 ± 0.4	21.6 ± 1.3	94 ± 8	424 ± 33	1.66 ± 0.14	90 ± 2	55 ± 3	50 ± 4
Recovery (%)	95 ± 8	103 ± 6	93 ± 8	95 ± 7	90 ± 8	95 ± 2	109 ± 5	100 ± 7

[a] Experimental conditions used: (i) sample weight, 10 g; (ii) sample pH, 7.6; (iii) amount of sorbent, 10 mg; (iv) dispersion mode, vortex; (v) complexing agent concentration, 0.5%; (vi) ionic concentration, 0%; (vii) eluent, HCl; (viii) eluent concentration, 0.5 M; (ix) eluent volume, 300 µL; (x) extraction time, 3 min; (xi) elution time, 3 min. [b] Certified value ± expanded uncertainty. [c] Found value ± expanded uncertainty (coverage factor of 2).

3.5. Method Applicability

Real wastewater sample was analyzed and recovery studies were carried out to assess the applicability of the proposed method. Real sample was spiked at two different levels (i.e., 5.05 and 25.4 $\mu g\ L^{-1}$) and analyzed in triplicate. All the measurements of the real sample were performed under optimal conditions. As can be seen in Table 3, the results show recovery values ranging between 94 and 108%, with standard deviations from 2 to 8% demonstrating that the developed method is valid for applications in performing analyses of real wastewater samples.

Table 3. Results of recovery study on spiked real sample [a].

	Added Value (µg L^{-1})	Analyte										
		Ba	Be	Cd	Cr	Cu	Fe	Hg	Mn	Ni	Pb	Zn
Found value (µg L^{-1}) [b]	-	<LOQ	<LOQ	<LOQ	308.4 ± 0.5	53.6 ± 1.0	62.6 ± 1.3	1.27 ± 0.05	2.41 ± 0.02	<LOQ	<LOQ	108.4 ± 1.4
	5.05	5.4 ± 0.2 (106 ± 4)	5.2 ± 0.2 (101 ± 6)	5.02 ± 0.16 (99 ± 3)	313.8 ± 0.4 (108 ± 8)	58.4 ± 0.3 (95 ± 6)	67.6 ± 0.2 (98 ± 4)	6.1 ± 0.2 (94 ± 3)	7.2 ± 0.2 (94 ± 5)	5.2 ± 0.2 (103 ± 4)	5.2 ± 0.2 (103 ± 4)	113.4 ± 0.3 (98 ± 6)
	25.4	26.1 ± 1.1 (102 ± 5)	24.0 ± 0.9 (94 ± 3)	24.0 ± 0.8 (94 ± 4)	334.2 ± 1.4 (101 ± 5)	79.6 ± 1.9 (102 ± 6)	88.8 ± 0.7 (103 ± 3)	27.0 ± 1.4 (100 ± 7)	26.5 ± 0.8 (95 ± 3)	27.2 ± 0.7 (107 ± 2)	27.2 ± 1.3 (107 ± 4)	133.4 ± 0.8 (98 ± 4)

[a] Experimental conditions used: (i) sample weight, 10 g; (ii) sample pH, 7.6; (iii) amount of sorbent, 10 mg; (iv) dispersion mode, vortex; (v) complexing agent concentration, 0.5%; (vi) ionic concentration, 0%; (vii) eluent, HCl; (viii) eluent concentration, 0.5 M; (ix) eluent volume, 300 µL; (x) extraction time, 3 min; (xi) elution time, 3 min. [b] Found value ± standard deviation. Recovery (%) ± standard deviation is shown in parentheses.

3.6. Comparison with Other Methods

A comparison between this work and previously described procedures that combined different SPE techniques and spectrochemical detection methods for the determination of trace metals in water and environmental samples is shown in Table S6. As can be seen, the developed method has the shortest extraction time (i.e., 3 min) and the lowest sorbent and sample amounts (i.e., 10 mg and 10 g, respectively) compared with previous studies. This is a great advantage according to the second (i.e., minimal sample size and minimal number of samples are goals), fifth (i.e., automated and miniaturized methods should be selected), and ninth (i.e., the use of energy should be minimized) principles of Green Analytical Chemistry (GAC) [85]. In addition, the developed method in this work has been evaluated for use with a higher number of analytes compared with the rest of the experimental studies; this is in accordance with the eighth principle of GAC (i.e., multi-analyte or multi-parameter methods are preferred versus methods using one analyte at a time). In some cases, the differences are outstanding: for example, in the method developed by Dogru et al. [1], a duration of at least 85 min was needed in order to prepare the column and perform the solid-phase extraction. In addition, the sorbent amount required in the proposed method was found to be between 2 and 25 times lower than other methods presented in the literature (Table S6), reducing the consumption of reagents and the cost of the MDSPE procedure. On the other hand, this work provides LOD values (i.e., 0.073–0.22 μg L^{-1}) that are lower or comparable with previous ICP OES-based studies; except for Fe, Ni, and Pb, where the limits of detection were 1.3, 1.1, and 1.1 μg L^{-1}, respectively.

4. Conclusions

The developed MDSPE-ICP OES method has been successfully applied for the preconcentration, separation, and determination of trace metals (i.e., Ba, Be, Cd, Cr, Cu, Fe, Hg, Mn, Ni, Pb, and Zn) in wastewater samples. The proposed sorbent is a new carbon material that contains magnetic cobalt; it was prepared using controlled pyrolysis of a ZIF-67. This magnetic material combines the excellent extraction properties of zeolites with the simple handling of the magnetic sorbent in the presence of an external magnetic field. Using this method, high enrichment factors (i.e., between 3.2 and 13) and lower limits of detection and quantification have been obtained compared with those obtained by direct ICP OES analysis. The limits of detection obtained in the developed method range from 0.073 to 1.1 μg L^{-1}; these values are, on average, 17 times lower than the values established in the official method of the Environmental Protection Agency (i.e., method 200.7). In addition, the MDSPE-ICP OES has been validated through the analysis of a certified reference material; this analysis obtained recovery values ranging from 90 to 109% of the certified value. Furthermore, the applicability of the method was evaluated using recovery experiments at two different spiking levels (i.e., recovery values from 94 to 108%) in a real wastewater sample. The magnetic nature of the sorbent material enables the separation through the application of an external magnetic field; this permits the centrifugation step to be avoided and reduces the consumption of energy and time used in the analysis. In addition, the proposed method has been developed for a high number of elements, and the sample and sorbent amounts required have been reduced compared with previous studies; these difference meet the requirements outlined by the GAC principles and ensure that this method is an ecofriendly option. Moreover, the synthesis of the magnetic sorbent is inexpensive, and can be disposed of after a single use. For all these reasons, the developed method is a fast, reliable, multi-elemental, and ecofriendly alternative to conventional SPE methods for use in the analysis of trace metals in wastewater samples. Furthermore, taking into account all the extraction properties and the low synthesis cost, the developed magnetic material and the new analytical method could comprise a potential option to perform the extraction of analytes in industrial wastewater treatment in comparison with current materials.

Supplementary Materials: The following supporting information can be downloaded at: https://www.mdpi.com/article/10.3390/separations10110563/s1, Figure S1: XRD pattern of ZIF67c_900_l; Figure S2: N2 adsorption isotherm of the original carbon (i.e., ZIF67c_900) and the washed material ZIF67c_900_l; Figure S3: XPS spectra of cobalt and nitrogen in ZIF67c_900_l; Figure S4: Raman spectrum of ZIF67c_900_l; Figure S5: Magnetization curve of ZIF67c_900_l: (a) representation of the magnetization curve (i.e., X axis from −25,000 to 25,000 Oe); (b) amplified representation of the magnetization curve (i.e., X axis from −2000 to 2000 Oe); Figure S6: Zeta potential of ZIF67c_900_l/DDTC complex as function of the pH value; Figure S7: Pareto charts obtained in the screening study of the factors affecting the MDSPE. Orange means significant effect while blue means non-significant effect on MDSPE; Figure S8: Response surface from CCD for each analyte evaluated; Table S1: Operational parameters employed in Agilent 720 ICP OES; Table S2: Considered experimental factors and levels in the Plackett–Burman design; Table S3: Considered experimental levels in CCD design; Table S4: Optimum sample pH and eluent volume for each determined element; Table S5: Optimum experimental conditions for MDSPE; Table S6: Comparison of analytical figures of merit of the proposed method and other published methods.

Author Contributions: Conceptualization, A.C. and J.N.; Methodology, D.V.-H., E.V.R.-F., S.J.A.-M. and M.Á.A.; Software, S.J.A.-M., D.V.-H. and M.Á.A.; Validation, S.J.A.-M., D.V.-H., E.V.R.-F. and M.Á.A.; Formal Analysis, S.J.A.-M. and D.V.-H.; Investigation, D.V.-H., E.V.R.-F., S.J.A.-M. and M.Á.A.; Resources, A.C. and J.N.; Data Curation, D.V.-H., E.V.R.-F., S.J.A.-M. and M.Á.A.; Writing—Original Draft Preparation, D.V.-H. and S.J.A.-M.; Writing—Review and Editing, D.V.-H., E.V.R.-F., J.N., S.J.A.-M., M.Á.A. and A.C.; Visualization, D.V.-H., E.V.R.-F., J.N., S.J.A.-M., M.Á.A. and A.C.; Supervision, A.C. and J.N.; Project Administration, A.C. and J.N.; Funding Acquisition, A.C. and J.N. All authors have read and agreed to the published version of the manuscript.

Funding: This research was funded by the Ministry of Spain grant number (PID2021-126155OB-I00, and PID2020-116998RB-I00) and by the Regional Government of Valencia (Spain) grant number (CIPROM/2021/062, CIPROM/2021/022, INVEST/2022/70, and PRTR-C17.I1).

Data Availability Statement: The data presented in this study are available on request from the corresponding authors.

Acknowledgments: The authors are grateful to Ingeniatrics for the OneNeb® Series 2 provided. This article is based upon work from the Sample Preparation Study Group and Network, supported by the Division of Analytical Chemistry of the European Chemical Society.

Conflicts of Interest: The authors declare no conflict of interest.

References

1. Dogru, M.; Gul-Guven, R.; Erdogan, S. The Use of Bacillus Subtilis Immobilized on Amberlite XAD-4 as a New Biosorbent in Trace Metal Determination. *J. Hazard. Mater.* **2007**, *149*, 166–173. [CrossRef] [PubMed]
2. Zouboulis, A.I.; Loukidou, M.X.; Matis, K.A. Biosorption of Toxic Metals from Aqueous Solutions by Bacteria Strains Isolated from Metal-Polluted Soils. *Process Biochem.* **2004**, *39*, 909–916. [CrossRef]
3. Bermejo-Barrera, P.; Martínez Alfonso, N.; Díaz López, C.; Bermejo Barrera, A. Use of Amberlite XAD-2 Loaded with 1-(2-Pyridylazo)-2-Naphthol as a Preconcentration System for River Water Prior to Determination of Cu^{2+}, Cd^{2+} and Pb^{2+} by Flame Atomic Absorption Spectroscopy. *Microchim. Acta* **2003**, *142*, 101–108. [CrossRef]
4. Sardans, J.; Montes, F.; Peñuelas, J. Determination of As, Cd, Cu, Hg and Pb in Biological Samples by Modern Electrothermal Atomic Absorption Spectrometry. *Spectrochim. Acta B* **2010**, *65*, 97–112. [CrossRef]
5. Wang, J.; Chen, C. Biosorbents for Heavy Metals Removal and Their Future. *Biotechnol. Adv.* **2009**, *27*, 195–226. [CrossRef]
6. Giakisikli, G.; Anthemidis, A.N. Magnetic Materials as Sorbents for Metal/Metalloid Preconcentration and/or Separation. A Review. *Anal. Chim. Acta* **2013**, *789*, 1–16. [CrossRef]
7. Harvey, P.J.; Handley, H.K.; Taylor, M.P. Identification of the Sources of Metal (Lead) Contamination in Drinking Waters in North-Eastern Tasmania Using Lead Isotopic Compositions. *Environ. Sci. Pollut. Res.* **2015**, *22*, 12276–12288. [CrossRef]
8. Ma, Y.; Egodawatta, P.; McGree, J.; Liu, A.; Goonetilleke, A. Human Health Risk Assessment of Heavy Metals in Urban Stormwater. *Sci. Total Environ.* **2016**, *557–558*, 764–772. [CrossRef] [PubMed]
9. Zhou, L.; Gao, C.; Xu, W. Magnetic Dendritic Materials for Highly Efficient Adsorption of Dyes and Drugs. *ACS Appl. Mater. Interfaces* **2010**, *2*, 1483–1491. [CrossRef]
10. Bagheri, H.; Afkhami, A.; Saber-Tehrani, M.; Khoshsafar, H. Preparation and Characterization of Magnetic Nanocomposite of Schiff Base/Silica/Magnetite as a Preconcentration Phase for the Trace Determination of Heavy Metal Ions in Water, Food and Biological Samples Using Atomic Absorption Spectrometry. *Talanta* **2012**, *97*, 87–95. [CrossRef]

11. Suleiman, J.S.; Hu, B.; Huang, C.; Zhang, N. Determination of Cd, Co, Ni and Pb in Biological Samples by Microcolumn Packed with Black Stone (Pierre Noire) Online Coupled with ICP-OES. *J. Hazard. Mater.* **2008**, *157*, 410–417. [CrossRef] [PubMed]
12. Yamini, Y.; Faraji, M.; Shariati, S.; Hassani, R.; Ghambarian, M. On-Line Metals Preconcentration and Simultaneous Determination Using Cloud Point Extraction and Inductively Coupled Plasma Optical Emission Spectrometry in Water Samples. *Anal. Chim. Acta* **2008**, *612*, 144–151. [CrossRef] [PubMed]
13. Davis, A.C.; Calloway, C.P.; Jones, B.T. Direct Determination of Cadmium in Urine by Tungsten-Coil Inductively Coupled Plasma Atomic Emission Spectrometry Using Palladium as a Permanent Modifier. *Talanta* **2007**, *71*, 1144–1149. [CrossRef]
14. Huang, C.; Hu, B. Silica-Coated Magnetic Nanoparticles Modified with γ-Mercaptopropyltrimethoxysilane for Fast and Selective Solid Phase Extraction of Trace Amounts of Cd, Cu, Hg, and Pb in Environmental and Biological Samples Prior to Their Determination by Inductively Co. *Spectrochim. Acta B* **2008**, *63*, 437–444. [CrossRef]
15. Zheng, F.; Hu, B. Thermo-Responsive Polymer Coated Fiber-in-Tube Capillary Microextraction and Its Application to on-Line Determination of Co, Ni and Cd by Inductively Coupled Plasma Mass Spectrometry (ICP-MS). *Talanta* **2011**, *85*, 1166–1173. [CrossRef]
16. Zhang, L.; Chen, B.; Peng, H.; He, M.; Hu, B. Aminopropyltriethoxysilane-Silica Hybrid Monolithic Capillary Microextraction Combined with Inductively Coupled Plasma Mass Spectrometry for the Determination of Trace Elements in Biological Samples. *J. Sep. Sci.* **2011**, *34*, 2247–2254. [CrossRef]
17. Naghizadeh, M.; Taher, M.A.; Behzadi, M.; Moghaddam, F.H. Preparation a Novel Magnetic Natural Nano Zeolite for Preconcentration of Cadmium and Its Determination by ETAAS. *Environ. Nanotechnol. Monit. Manag.* **2017**, *8*, 261–267. [CrossRef]
18. Jorhem, L.; Åstrand, C.; Sundström, B.; Baxter, M.; Stokes, P.; Lewis, J.; Grawe, K.P. Elements in Rice from the Swedish Market: 1. Cadmium, Lead and Arsenic (Total and Inorganic). *Food Addit. Contam.-Part A* **2008**, *25*, 284–292. [CrossRef]
19. Montoro-Leal, P.; García-Mesa, J.C.; Siles Cordero, M.T.; López Guerrero, M.M.; Vereda Alonso, E. Magnetic Dispersive Solid Phase Extraction for Simultaneous Enrichment of Cadmium and Lead in Environmental Water Samples. *Microchem. J.* **2020**, *155*, 104796. [CrossRef]
20. Xie, F.; Lin, X.; Wu, X.; Xie, Z. Solid Phase Extraction of Lead (II), Copper (II), Cadmium (II) and Nickel (II) Using Gallic Acid-Modified Silica Gel Prior to Determination by Flame Atomic Absorption Spectrometry. *Talanta* **2008**, *74*, 836–843. [CrossRef]
21. Tuzen, M.; Soylak, M. Column Solid-Phase Extraction of Nickel and Silver in Environmental Samples Prior to Their Flame Atomic Absorption Spectrometric Determinations. *J. Hazard. Mater.* **2009**, *164*, 1428–1432. [CrossRef] [PubMed]
22. Ghaedi, M.; Tavallali, H.; Shokrollahi, A.; Zahedi, M.; Montazerozohori, M.; Soylak, M. Flame Atomic Absorption Spectrometric Determination of Zinc, Nickel, Iron and Lead in Different Matrixes after Solid Phase Extraction on Sodium Dodecyl Sulfate (SDS)-Coated Alumina as Their Bis (2-Hydroxyacetophenone)-1, 3-Propanediimine Chelates. *J. Hazard. Mater.* **2009**, *166*, 1441–1448. [CrossRef] [PubMed]
23. Bryce, D.W.; Izquierdo, A.; Luque De Castro, M.D. Continuous Microwave Assisted Pervaporation/Atomic Fluorescence Detection: An Approach for Speciation in Solid Samples. *Anal. Chim. Acta* **1996**, *324*, 69–75. [CrossRef]
24. Zhang, W.; Su, Z.; Chu, X.; Yang, X. Evaluation of a New Electrolytic Cold Vapor Generation System for Mercury Determination by AFS. *Talanta* **2010**, *80*, 2106–2112. [CrossRef] [PubMed]
25. Cheng, G.; He, M.; Peng, H.; Hu, B. Dithizone Modified Magnetic Nanoparticles for Fast and Selective Solid Phase Extraction of Trace Elements in Environmental and Biological Samples Prior to Their Determination by ICP-OES. *Talanta* **2012**, *88*, 507–515. [CrossRef] [PubMed]
26. Zeini Jahromi, E.; Bidari, A.; Assadi, Y.; Milani Hosseini, M.R.; Jamali, M.R. Dispersive Liquid-Liquid Microextraction Combined with Graphite Furnace Atomic Absorption Spectrometry. Ultra Trace Determination of Cadmium in Water Samples. *Anal. Chim. Acta* **2007**, *585*, 305–311. [CrossRef]
27. Shabani, A.M.H.; Dadfarnia, S.; Nasirizadeh, N.; Shishehbore, M.R. Preconcentration of Copper with Dithizone-Naphthalene for Subsequent Determination by Atomic Absorption Spectrometry. *J. Anal. Chem.* **2007**, *62*, 46–50. [CrossRef]
28. Miró, M.; Hansen, E.H. Recent Advances and Future Prospects of Mesofluidic Lab-on-a-Valve Platforms in Analytical Sciences—A Critical Review. *Anal. Chim. Acta* **2012**, *750*, 3–15. [CrossRef]
29. Baghdadi, M.; Shemirani, F. Cold-Induced Aggregation Microextraction: A Novel Sample Preparation Technique Based on Ionic Liquids. *Anal. Chim. Acta* **2008**, *613*, 56–63. [CrossRef]
30. Soylak, M.; Yilmaz, E. Ionic Liquid Dispersive Liquid-Liquid Microextraction of Lead as Pyrrolidinedithiocarbamate Chelate Prior to Its Flame Atomic Absorption Spectrometric Determination. *Desalination* **2011**, *275*, 297–301. [CrossRef]
31. Pena-Pereira, F.; Lavilla, I.; Bendicho, C. Miniaturized Preconcentration Methods Based on Liquid-Liquid Extraction and Their Application in Inorganic Ultratrace Analysis and Speciation: A Review. *Spectrochim. Acta B* **2009**, *64*, 1–15. [CrossRef]
32. Soylak, M.; Aydin, A. Determination of Some Heavy Metals in Food and Environmental Samples by Flame Atomic Absorption Spectrometry after Coprecipitation. *Food Chem. Toxicol.* **2011**, *49*, 1242–1248. [CrossRef] [PubMed]
33. Soylak, M.; Kars, A.; Narin, I. Coprecipitation of Ni^{2+}, Cd^{2+} and Pb^{2+} for Preconcentration in Environmental Samples Prior to Flame Atomic Absorption Spectrometric Determinations. *J. Hazard. Mater.* **2008**, *159*, 435–439. [CrossRef] [PubMed]
34. Saberyan, K.; Zolfonoun, E.; Shamsipur, M.; Salavati-Niasari, M. Separation and Preconcentration of Trace Gallium and Indium by Amberlite XAD-7 Resin Impregnated with a New Hexadentates Naphthol-Derivative Schiff Base. *Sep. Sci. Technol.* **2009**, *44*, 1851–1868. [CrossRef]

35. Uludag, Y.; Özbelge, H.Ö.; Yilmaz, L. Removal of Mercury from Aqueous Solutions via Polymer-Enhanced Ultrafiltration. *J. Membr. Sci.* **1997**, *129*, 93–99. [CrossRef]
36. Bruno, P.; Caselli, M.; De Gennaro, G.; Ielpo, P.; Ladisa, T.; Placentino, C.M. Ion Chromatography Determination of Heavy Metals in Airborne Particulate with Preconcentration and Large Volume Direct Injection. *Chromatographia* **2006**, *64*, 537–542. [CrossRef]
37. Dą browski, A.; Hubicki, Z.; Podkościelny, P.; Robens, E. Selective Removal of the Heavy Metal Ions from Waters and Industrial Wastewaters by Ion-Exchange Method. *Chemosphere* **2004**, *56*, 91–106. [CrossRef]
38. Sánchez Rojas, F.; Ojeda, C.B.; Pavón, J.M.C. On-Line Preconcentration of Rhodium on an Anion-Exchange Resin Loaded with 1,5-Bis(2-Pyridyl)-3-Sulphophenyl Methylene Thiocarbonohydrazide and Its Determination in Environmental Samples. *Talanta* **2004**, *64*, 230–236. [CrossRef]
39. Shariati, S.; Yamini, Y.; Zanjani, M.K. Simultaneous Preconcentration and Determination of U(VI), Th(IV), Zr(IV) and Hf(IV) Ions in Aqueous Samples Using Micelle-Mediated Extraction Coupled to Inductively Coupled Plasma-Optical Emission Spectrometry. *J. Hazard. Mater.* **2008**, *156*, 583–590. [CrossRef]
40. Meeravali, N.N.; Jiang, S.J. Microwave Assisted Mixed-Micelle Cloud Point Extraction of Au and Tl from Environmental Samples without Using a Chelating Agent Prior to ICP-MS Determination. *J. Anal. At. Spectrom.* **2008**, *23*, 1365–1371. [CrossRef]
41. Tuzen, M.; Karaman, I.; Citak, D.; Soylak, M. Mercury(II) and Methyl Mercury Determinations in Water and Fish Samples by Using Solid Phase Extraction and Cold Vapour Atomic Absorption Spectrometry Combination. *Food Chem. Toxicol.* **2009**, *47*, 1648–1652. [CrossRef] [PubMed]
42. Tuzen, M.; Uluozlu, O.D.; Karaman, I.; Soylak, M. Mercury(II) and Methyl Mercury Speciation on Streptococcus Pyogenes Loaded Dowex Optipore SD-2. *J. Hazard. Mater.* **2009**, *169*, 345–350. [CrossRef] [PubMed]
43. Fan, J.; Qin, Y.; Ye, C.; Peng, P.; Wu, C. Preparation of the Diphenylcarbazone-Functionalized Silica Gel and Its Application to on-Line Selective Solid-Phase Extraction and Determination of Mercury by Flow-Injection Spectrophotometry. *J. Hazard. Mater.* **2008**, *150*, 343–350. [CrossRef] [PubMed]
44. Faraji, M.; Yamini, Y.; Rezaee, M. Extraction of Trace Amounts of Mercury with Sodium Dodecyle Sulphate-Coated Magnetite Nanoparticles and Its Determination by Flow Injection Inductively Coupled Plasma-Optical Emission Spectrometry. *Talanta* **2010**, *81*, 831–836. [CrossRef]
45. Poole, C.F. New Trends in Solid-Phase Extraction. *TrAC-Trends Anal. Chem.* **2003**, *22*, 362–373. [CrossRef]
46. Zhang, C.; Xing, H.; Yang, L.; Fei, P.; Liu, H. Development Trend and Prospect of Solid Phase Extraction Technology. *Chin. J. Chem. Eng.* **2022**, *42*, 245–255. [CrossRef]
47. Giakisikli, G.; Anthemidis, A.N. Automated Magnetic Sorbent Extraction Based on Octadecylsilane Functionalized Maghemite Magnetic Particles in a Sequential Injection System Coupled with Electrothermal Atomic Absorption Spectrometry for Metal Determination. *Talanta* **2013**, *110*, 229–235. [CrossRef]
48. Rao, T.P.; Metilda, P.; Gladis, J.M. Preconcentration Techniques for Uranium(VI) and Thorium(IV) Prior to Analytical Determination-an Overview. *Talanta* **2006**, *68*, 1047–1064. [CrossRef]
49. Zhu, F.; Ma, Z.; Qiu, K.; Peng, W. Separation of Ilmenite from Vanadium Titanomagnetite by Combining Magnetic Separation and Flotation Processes. *Separations* **2023**, *10*, 95. [CrossRef]
50. Córdova Udaeta, M.; Dodbiba, G.; Ponou, J.; Sone, K.; Fujita, T. Recovery of Phosphorus from Sewage Sludge Ash (SSA) by Heat Treatment Followed by High Gradient Magnetic Separation and Flotation. *Adv. Powder Technol.* **2017**, *28*, 755–762. [CrossRef]
51. Jiménez-Soto, J.M.; Cárdenas, S.; Valcárcel, M. Evaluation of Single-Walled Carbon Nanohorns as Sorbent in Dispersive Micro Solid-Phase Extraction. *Anal. Chim. Acta* **2012**, *714*, 76–81. [CrossRef] [PubMed]
52. Galán-Cano, F.; Lucena, R.; Cárdenas, S.; Valcárcel, M. Direct Coupling of Dispersive Micro-Solid Phase Extraction and Thermal Desorption for Sensitive Gas Chromatographic Analysis. *Anal. Methods* **2011**, *3*, 991–995. [CrossRef]
53. Skorek, R.; Zawisza, B.; Marguí, E.; Queralt, I.; Sitko, R. Dispersive Micro Solid-Phase Extraction Using Multiwalled Carbon Nanotubes for Simultaneous Determination of Trace Metal Ions by Energy-Dispersive x-Ray Fluorescence Spectrometry. *Appl. Spectrosc.* **2013**, *67*, 204–209. [CrossRef] [PubMed]
54. Ramutshatsha-Makhwedzha, D.; Mbaya, R.; Mavhungu, M.L.; Nomngongo, P.N. Ultrasonic Assisted Dispersive-Solid Phase Extraction for Preconcentration of Trace Metals in Wastewater Samples. *J. Environ. Chem. Eng.* **2022**, *10*, 108187. [CrossRef]
55. Armenta, S.; Garrigues, S.; de la Guardia, M. Green Analytical Chemistry. *TrAC-Trends Anal. Chem.* **2008**, *27*, 497–511. [CrossRef]
56. Duan, C.; Shen, Z.; Wu, D.; Guan, Y. Recent Developments in Solid-Phase Microextraction for on-Site Sampling and Sample Preparation. *TrAC-Trends Anal. Chem.* **2011**, *30*, 1568–1574. [CrossRef]
57. Socas-Rodríguez, B.; Herrera-Herrera, A.V.; Asensio-Ramos, M.; Hernández-Borges, J. Dispersive Solid-Phase Extraction. In *Analytical Separation Science*; Wiley-VCH Verlag GmbH & Co. KGaA: Weinheim, Germany, 2015; pp. 1525–1570. [CrossRef]
58. Ruiz, F.J.; Ripoll, L.; Hidalgo, M.; Canals, A. Dispersive Micro Solid-Phase Extraction (DμSPE) with Graphene Oxide as Adsorbent for Sensitive Elemental Analysis of Aqueous Samples by Laser Induced Breakdown Spectroscopy (LIBS). *Talanta* **2019**, *191*, 162–170. [CrossRef]
59. Lord, H.; Pawliszyn, J. Evolution of Solid-Phase Microextraction Technology. *J. Chromatogr. A* **2000**, *885*, 153–193. [CrossRef]
60. Kharissova, O.V.; Dias, H.V.R.; Kharisov, B.I. Magnetic Adsorbents Based on Micro- and Nano-Structured Materials. *RSC Adv.* **2015**, *5*, 6695–6719. [CrossRef]
61. Tian, J.; Xu, J.; Zhu, F.; Lu, T.; Su, C.; Ouyang, G. Application of Nanomaterials in Sample Preparation. *J. Chromatogr. A* **2013**, *1300*, 2–16. [CrossRef]

62. Płotka-Wasylka, J.; Szczepańska, N.; de la Guardia, M.; Namieśnik, J. Miniaturized Solid-Phase Extraction Techniques. *TrAC-Trends Anal. Chem.* **2015**, *73*, 19–38. [CrossRef]
63. Wierucka, M.; Biziuk, M. Application of Magnetic Nanoparticles for Magnetic Solid-Phase Extraction in Preparing Biological, Environmental and Food Samples. *TrAC-Trends Anal. Chem.* **2014**, *59*, 50–58. [CrossRef]
64. Safarikova, M.; Kibrikova, I.; Ptackova, L.; Hubka, T.; Komarek, K.; Safarik, I. Magnetic Solid Phase Extraction of Non-Ionic Surfactants from Water. *J. Magn. Magn. Mater.* **2005**, *293*, 377–381. [CrossRef]
65. Aliyari, E.; Alvand, M.; Shemirani, F. Modified Surface-Active Ionic Liquid-Coated Magnetic Graphene Oxide as a New Magnetic Solid Phase Extraction Sorbent for Preconcentration of Trace Nickel. *RSC Adv.* **2016**, *6*, 64193–64202. [CrossRef]
66. Asfaram, A.; Ghaedi, M.; Goudarzi, A.; Soylak, M.; Mehdizadeh Langroodi, S. Magnetic Nanoparticle Based Dispersive Micro-Solid-Phase Extraction for the Determination of Malachite Green in Water Samples: Optimized Experimental Design. *New J. Chem.* **2015**, *39*, 9813–9823. [CrossRef]
67. Costa dos Reis, L.; Vidal, L.; Canals, A. Determination of Siloxanes in Water Samples Employing Graphene Oxide/Fe_3O_4 Nanocomposite as Sorbent for Magnetic Solid-Phase Extraction Prior to GC–MS. *J. Sep. Sci.* **2018**, *41*, 4177–4184. [CrossRef]
68. Aguilar-Arteaga, K.; Rodriguez, J.A.; Barrado, E. Magnetic Solids in Analytical Chemistry: A Review. *Anal. Chim. Acta* **2010**, *674*, 157–165. [CrossRef]
69. Chen, J.; Wang, Y.; Wei, X.; Xu, P.; Xu, W.; Ni, R.; Meng, J. Magnetic Solid-Phase Extraction for the Removal of Mercury from Water with Ternary Hydrosulphonyl-Based Deep Eutectic Solvent Modified Magnetic Graphene Oxide. *Talanta* **2018**, *188*, 454–462. [CrossRef]
70. Mollahosseini, A.; Toghroli, M.; Kamankesh, M. Zeolite/Fe_3O_4 as a New Sorbent in Magnetic Solid-Phase Extraction Followed by Gas Chromatography for Determining Phthalates in Aqueous Samples. *J. Sep. Sci.* **2015**, *38*, 3750–3757. [CrossRef]
71. Tolmacheva, V.V.; Apyari, V.V.; Kochuk, E.V.; Dmitrienko, S.G. Magnetic Adsorbents Based on Iron Oxide Nanoparticles for the Extraction and Preconcentration of Organic Compounds. *J. Anal. Chem.* **2016**, *71*, 321–338. [CrossRef]
72. Hemmati, M.; Rajabi, M.; Asghari, A. Magnetic Nanoparticle Based Solid-Phase Extraction of Heavy Metal Ions: A Review on Recent Advances. *Microchim. Acta* **2018**, *185*, 160. [CrossRef] [PubMed]
73. Baile, P.; Vidal, L.; Canals, A. A Modified Zeolite/Iron Oxide Composite as a Sorbent for Magnetic Dispersive Solid-Phase Extraction for the Preconcentration of Nonsteroidal Anti-Inflammatory Drugs in Water and Urine Samples. *J. Chromatogr. A* **2019**, *1603*, 33–43. [CrossRef]
74. Baile, P.; Fernández, E.; Vidal, L.; Canals, A. Zeolites and Zeolite-Based Materials in Extraction and Microextraction Techniques. *Analyst* **2019**, *144*, 366–387. [CrossRef] [PubMed]
75. Baile, P.; Vidal, L.; Canals, A. Magnetic Dispersive Solid-Phase Extraction Using ZSM-5 Zeolite/Fe_2O_3 Composite Coupled with Screen-Printed Electrodes Based Electrochemical Detector for Determination of Cadmium in Urine Samples. *Talanta* **2020**, *220*, 121394. [CrossRef]
76. Villalgordo-Hernández, D.; Grau-Atienza, A.; García-Marín, A.A.; Ramos-Fernández, E.V.; Narciso, J. Manufacture of Carbon Materials with High Nitrogen Content. *Materials* **2022**, *15*, 2415. [CrossRef]
77. Zabiegaj, D.; Caccia, M.; Casco, M.E.; Ravera, F.; Narciso, J. Synthesis of Carbon Monoliths with a Tailored Hierarchical Pore Structure for Selective CO_2 Capture. *J. CO2 Util.* **2018**, *26*, 36–44. [CrossRef]
78. Narciso, J.; Ramos-Fernandez, E.V.; Delgado-Marín, J.J.; Affolter, C.W.; Olsbye, U.; Redekop, E.A. New Route for the Synthesis of Co-MOF from Metal Substrates. *Microporous Mesoporous Mater.* **2021**, *324*, 111310. [CrossRef]
79. Ramos-Fernandez, E.V.; Redondo-Murcia, A.; Grau-Atienza, A.; Sepúlveda-Escribano, A.; Narciso, J. Clean Production of Zeolitic Imidazolate Framework 8 Using Zamak Residues as Metal Precursor and Substrate. *J. Clean. Prod.* **2020**, *260*, 121081. [CrossRef]
80. Delgado-Marín, J.J.; Martín-García, I.; Villalgordo-Hernández, D.; Alonso, F.; Ramos-Fernández, E.V.; Narciso, J. Valorization of CO_2 through the Synthesis of Cyclic Carbonates Catalyzed by ZIFs. *Molecules* **2022**, *27*, 7791. [CrossRef]
81. Caccia, M.; Giuranno, D.; Molina-Jorda, J.M.; Moral, M.; Nowak, R.; Ricci, E.; Sobczak, N.; Narciso, J.; Fernández Sanz, J. Graphene Translucency and Interfacial Interactions in the Gold/Graphene/SiC System. *J. Phys. Chem. Lett.* **2018**, *9*, 3850–3855. [CrossRef]
82. Costa, D.A.S.; Mambrini, R.V.; Fernandez-Outon, L.E.; Macedo, W.A.A.; Moura, F.C.C. Magnetic Adsorbent Based on Cobalt Core Nanoparticles Coated with Carbon Filaments and Nanotubes Produced by Chemical Vapor Deposition with Ethanol. *Chem. Eng. J.* **2013**, *229*, 35–41. [CrossRef]
83. Baile, P.; Vidal, L.; Aguirre, M.Á.; Canals, A. A Modified ZSM-5 Zeolite/Fe_2O_3 Composite as a Sorbent for Magnetic Dispersive Solid-Phase Microextraction of Cadmium, Mercury and Lead from Urine Samples Prior to Inductively Coupled Plasma Optical Emission Spectrometry. *J. Anal. At. Spectrom.* **2018**, *33*, 856–866. [CrossRef]
84. Magnusson, B.; Örnemark, U. *Eurachem Guide: The Fitness for Purpose of Analytical Methods—A Laboratory Guide to Method Validation and Related Topics*, 2nd ed.; LGC: Teddington Ltd.: Teddington, UK, 2014; ISBN 9789187461590.
85. Gałuszka, A.; Migaszewski, Z.; Namieśnik, J. The 12 Principles of Green Analytical Chemistry and the SIGNIFICANCE Mnemonic of Green Analytical Practices. *TrAC-Trends Anal. Chem.* **2013**, *50*, 78–84. [CrossRef]

water

MDPI

Article

Concentration, Spatial Distribution, and Source Analysis of Trace Elements in the Yarlung Zangbo River Basin and Its Two Tributaries

Fangjing Xiao, Yuanzhao Zhao, Duo Bu and Qingying Zhang *

College of Ecological Environment, Tibet University, Lhasa 850000, China; xfj2023@utibet.edu.cn (F.X.); 15617927906@163.com (Y.Z.)
* Correspondence: phudor@vip.163.com (D.B.); zhangqiangying@utibet.edu.cn (Q.Z.)

Abstract: The Yarlung Zangbo River (YZR) is the longest plateau river in China and has famous tributaries, the Lhasa River and the Nianchu River. A total of 75 water samples were collected from the Yarlung Zangbo River Basin (YZRB) in this study to investigate the dissolved concentration, spatial distribution, and source of trace elements (Fe, V, Be, Ti, Mo, Se, Cd, Zn, Cu, Ni, Co, Mn, Cr, Ba, Tl, Pb, Hg, As, and Sb). The results indicate that only Cr and Tl contaminate water, while the other trace elements were in an unpolluted state. In addition, correlation analysis showed that there was a highly significant positive correlation between the concentrations of As, Sb, and Mo; there was also a highly significant positive correlation between the concentrations of Fe, Mn, Ti, Pb, Ni, Co, and Ba. The results of Positive Matrix Factorization (PMF) showed that there were four sources of trace elements in the YZRB, including the resuspension and dissolution of sediments (16.283%), agricultural source (11.436%), lithological source (47.418%), and soil-forming rocks (6.374%). Cluster analysis combined with PMF normalized contribution analysis, which showed that the trace elements found in the YZR's mainstream were predominantly influenced by the surrounding rocks composition. Meanwhile, both the discharge of mining wastewater and sediments were marked in the Lhasa River. Additionally, agricultural activities were the chief contributors to the trace elements in the Nianchu River. Furthermore, the entire basin was subjected to the influence of soil-forming rocks. This study comprehensively analyzed and evaluated the physicochemical properties of water, the spatial distribution, and the pollution degree, and performed source analysis of trace elements in the YZRB. This research provides a foundational reference for further investigation of the spatial distribution and origins of trace elements in the rivers of the Qinghai–Tibet Plateau (QTP).

Keywords: Yarlung Zangbo River; trace elements; spatial distribution; pollution degree; source analysis

1. Introduction

The Qinghai–Tibet Plateau (QTP), with its status as the "Asia water tower", is the main water source and ecological barrier in Asia. Being both the most crucial and fragile reservoir of water resources, the inherent significance of this water tower is undeniable and has been widely acknowledged in numerous authoritative studies [1,2]. Generally, the QTP was considered to be the most limited primitive area of human influence, where most of the rivers in Tibet were considered uncontaminated [3]. Nowadays, with the intensification of human activity (such as tourism, mining operations, urban sewage discharge, and arable land), the potential risks caused by trace elements to local ecosystems have gradually become prominent [4].

Heavy metal pollution in the QTP has appeared in different environmental media, especially in some rivers [5]. Although the rivers in the QTP can usually be regarded as primitive rivers, due to the increasing impact of human activity, the concentration of some toxic elements (i.e., Cd and Pb) is higher than the World Health Organization's drinking water guidelines, which poses a risk to human livelihoods in the local and surrounding areas. Wang et al. stated that some trace elements, such as Cu (13.92 μg/L), As (41.60 μg/L), and Pb (26.69 μg/L) in the lake of Yamdrok–tso Basin were characterized by high dissolved concentration values [6]. Zhang et al. revealed that the significant enrichment of As in Singe Zangbo River water (up to 104 μg/L) was sourced from hot springs and evaporite [7].

Trace elements enter the water through mining wastewater [8], the chemical industry [9], pesticides and fertilizers [10], atmospheric deposition [11], and soil weathering [12]. They are important factors affecting the safety evaluation of the water's ecological environment. Therefore, Fe, V, Be, Ti, Mo, Se, Cd, Zn, Cu, Ni, Co, Mn, Cr, Ba, Tl, Pb, Hg, As, and Sb were used as water quality monitoring indicators to assess the Environmental Quality Standards for Surface Water (GB3838–2002) [13]. To comprehensively understand the impact of trace elements on water quality and health risks, it is usually necessary to classify the sources of trace elements and use a pollution index and health risk index to describe the degree of heavy metal pollution and health hazards. Therefore, studying the water quality status and health risks of the Tibet Plateau is of crucial significance when revealing the potential threats to water security and protecting water resources. The research results of trace elements in the water of the study basin are not comprehensive enough. Therefore, this paper analyzes the spatial distribution characteristics and source analysis of trace elements in different reaches of the Yarlung Zangbo River Basin (YZRB).

The purpose of this study is (1) to reveal the concentration and spatial distribution characteristics of trace elements in the water of the YZRB and its two tributaries; (2) to evaluate the pollution characteristics of trace elements in the watershed; (3) to trace its possible sources and analyze its proportion using the Positive Matrix Factorization (PMF). This comprehensive study thoroughly analyzes and assesses the physicochemical properties, trace element pollution characteristics, spatial distribution patterns, and source apportionment within the YZRB. It unveils the current water environmental quality of the YZRB and serves as a valuable reference for investigating the concentration, distribution characteristics, and sources of various trace elements in rivers across the Tibetan Plateau.

2. Materials and Methods

2.1. Study Area

The study area (28°40′44″ N~29°51′38″ N, 88°51′50″ E~92°45′34″ E) of YZRB is located in the southern part of the Tibet Plateau in China, including the middle of the YZRB and the two tributaries (Lhasa River and Nianchu River), a well-known base for self-evident food production. As shown in Figure 1, a tributary of the Lhasa River is located in the capital city of Tibet (Lhasa), exposed to copper mining and distributions of geothermal resources [14]. The most common pollutants entering the river are Cu and As. In contrast, the other tributary (Nianchu River) is densely populated with facility agriculture [15], indicating that the main source of pollution is agricultural non-point source pollution. Overall, the YZRB exhibits ancient karst geological characteristics [16]. From September to October 2021, water samples (n = 75) were collected and presented as Y1–20 (YZR's mainstream), L1–14 (Lhasa River), and N1–41 (Nianchu River). The average altitude of the sampling points reached 3762 m. The basin is also rich in species diversity and is affected by unique high-altitude cold climatic conditions, making it one of the most vulnerable areas in Asia.

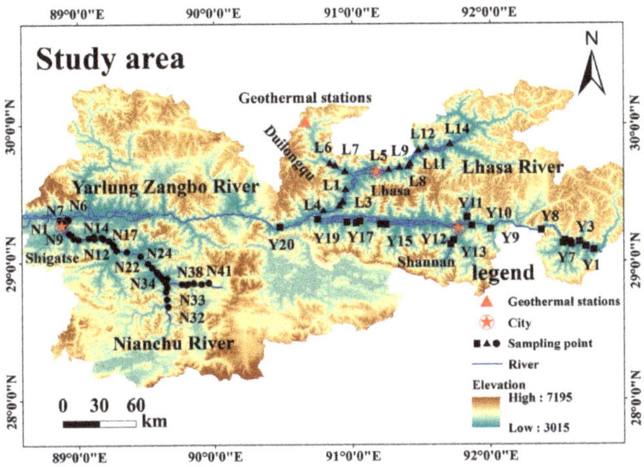

Figure 1. Study area and the geographical location of sampling sites.

2.2. Sample Collection and Analysis

According to the terrain and on-site environment, the spacing between each sampling site is about 2.5–10 km. Specifically, some important sampling sites (Y19, N6) are set after tributary flows into the mainstream and are also fully mixed. By comparison, the control section sampling sites (N7, L4) are set before tributary flows into the mainstream. The collection of water samples is conducted as follows: prior to sampling, the sampling apparatus, containers, and container lids were acid-washed with 10% (v/v) high-quality pure nitric acid. Water samples were collected in 500 mL acid-cleaned brown polyethylene bottles, 10 cm below the water surface. After multi-parameter analysis of the water sample on site, the collected sample was filtered through acid-cleaned Millipore filters (25 mm diameter, 0.45 μm mixed cellulose esters membrane). Approximately 15 mL filtered water samples were transferred into polyethylene bottles (pre-acid-washed with 10%, v/v, high-quality pure nitric acid) and stored in boxes with ice. After collection, the water samples were transported to the laboratory for trace elements analysis. To determine the field parameters, a portable dissolved oxygen meter (HANNA, HI98193, Rhode Island, Italy) and a portable multi-parameter measuring instrument (Thermo, Eutech PCS Testr 35, Massachusetts, USA) were used to measure the dissolved oxygen (DO), temperature (T), electrical conductivity (EC), pH, total dissolved solids (TDS), and salinity (SAL) in water. In this study, the concentrations of Fe, V, Be, Ti, Mo, Se, Cd, Zn, Cu, Ni, Co, Mn, Cr, Ba, Tl, Pb, and Sb were detected using an inductively coupled plasma mass spectrometer (ICP-MS, PerkinElmer, NexION1000, Waltham, MA, USA). The concentrations of As and Hg were detected via an atomic fluorescence spectrometer (AFS, Beijing Titan, AFS-830, Beijing, China).

2.3. Evaluation of the Contamination

The Nemerow integrated pollution index method is a commonly used and representative method for the evaluation of trace elements in water [17]. It is divided into a single-factor pollution index (P_i) and a multi-factor-integrated pollution index (P_n). The evaluation of P_i can only reflect the pollution status of a single element, while P_n considers the influence of extreme values or prominent maximum values of multiple elements. Multiple trace elements often coexist in water, and the Nemerow-integrated pollution index method can reflect the status of trace element pollution in water and the different contributions various trace elements form in composite pollution. It also screens the main pollutants, which is also one of the common methods for the evaluation of trace element pollution in water.

$$P_i = C_i/S_i \tag{1}$$

$$P_n = \sqrt{\frac{max(P_i)^2 + ave(P_i)^2}{2}} \quad (2)$$

Here, C_i is the measured concentration of trace element i; S_i is the corresponding standard reference concentration, using the standard reference values of the Chinese standard GB3838–2002 [13]. Max (P_i) is the maximum value of the trace element single-factor pollution index, and ave (P_i) is the average value of the single-factor pollution index of trace elements. The classification of the potential contamination level of trace elements is presented in Table 1 [18].

Table 1. Classification of the potential contamination level of trace elements.

P_i	Pollution Grading	P_n	Pollution Grading
$P_i \leq 1$	no pollution	$P_n \leq 0.7$	clean level
$1 < P_i \leq 2$	low pollution	$0.7 < P_n \leq 1$	precaution level
$2 < P_i \leq 3$	moderate pollution	$1 < P_n \leq 2$	light pollution level
$3 < P_i \leq 5$	strong pollution	$2 < P_n \leq 3$	moderate level
$P_i > 5$	very strong pollution	$P_n > 3$	heavy pollution leve

2.4. Source Apportionment

PMF is a method for the quantitative analysis of pollution sources using sample composition [19]; it decomposes the pollutant concentration matrix X into a source factor contribution matrix G, factor distribution matrix F, and residual matrix E, which are then calculated as follows:

$$X_{ij} = \sum_{k=1}^{p} G_{ik}F_{kj} + E_{ij} \quad (3)$$

Here, X_{ij} is the measurement matrix of the jth trace element in the i sample; G_{ik} is the contribution matrix of the kth source factor of the i sample; F_{kj} is the distribution matrix of the jth trace element in the kth source factor; and E_{ij} is the residual value of the jth trace element in the i sample. The cumulative residual Q-value can be minimized for factor contribution and distribution using the PMF model.

$$Q = \sum_{i=1}^{n} \sum_{j=1}^{m} \left[\frac{X_{ij} - \sum_{k=1}^{p} G_{ik}F_{kj}}{U_{ij}} \right]^2 \quad (4)$$

Here, PMF is required to import uncertainty data before it is run. If the trace element concentration does not exceed the species-specific method detection limit (MDL value), then the value of uncertainty (U_{nc}) is equal to 5/6 MDL; otherwise, the uncertainty U_{nc} is calculated using Equation (6):

$$\text{If } X_{ij} \leq \text{MDL} \quad U_{nc} = 5/6 \times \text{MDL} \quad (5)$$

$$U_{nc} = \sqrt{(\text{Error Fraction} \times \text{concentration})^2 + (0.5 \times \text{MDL})^2} \quad (6)$$

Here, X_{ij} is the measurement matrix of the jth trace element in i samples, and the error fraction is the relative standard deviation of trace element concentrations.

2.5. Statistical Analysis

The sampling layout was carried out using statistics and ArcGIS 10.7, and the Kriging interpolation method revealed variability and spatial and temporal distribution characteristics of the trace elements. The Nemerow-integrated pollution index was used to analyze the contamination degree of the trace elements. PMF was used to discuss the correlation, source number, and contribution rate of trace elements in the YZRB and its two tributaries. PMF source apportionment was conducted using the matrix factorization model of EPA PMF 5.0. IBM SPSS Statistics 25 and Origin Pro 2021 software were used for systematic cluster analysis.

3. Results and Discussion

3.1. Physical and Chemical Parameters

Conventional physicochemical parameters characterize the basic water environment characteristics of YZRB. The conventional physicochemical parameters of water mainly include DO, T, EC, pH, TDS, and SAL. The average pH value of the basin was ranked as Nianchu River (8.92) > Lhasa River (8.42) > YZR's mainstream (7.32). Due to the fact that sampling was conducted during the warm season, the water temperature and DO values in the basin were relatively high at 14.7 °C and 4.61 mg/L, respectively. In addition, EC (mean 443 μS/cm), SAL (mean 214 mg/L), and TDS (mean 313 mg/L) in the YZRB were higher than the average value of the world river [20], which is consistent with the strong erosion of the plateau basin and the lithological composition of carbonate rocks. In addition, EC, TDS, and SAL showed a significant positive correlation ($p < 0.01$).

3.2. Evaluation of the Contamination

For comparison, the average concentrations of trace elements in the YZRB reported in the literature [3,21–26] were listed, as shown in Table 2. The trace elements (average concentration) measured in this work were found not to be significantly different from those reported in the literature, and the concentration levels were observed to be at almost the same level. Moreover, the reliability of the original data was tested using a standard solution, and it was found that the accuracy of the elements was between 74.5% (V) and 107% (Pb), indicating that the determination method is accurate and reliable.

Based on the concentration levels detected using trace elements, the single-factor (P_i) and multi-factor-integrated pollution index (P_n) of trace elements were evaluated. Among all the surveyed elements, elements with a P_i value less than one exhibited no pollution; therefore, these elements are not discussed. As shown in Figure 2a, only five trace elements (Cr, Cu, Fe, Tl, and Ti) have a maximum P_i value greater than one, indicating low pollution. Overall, the evaluation results of the P_i value indicate that YZRB was on a clean level; however, some sampling sites (2–7) showed outliers. Similar to the P_n value presented in Figure 2b, the elements of Cr and Tl had P_n values greater than 0.7, while the order of pollution degree was Tl (1.03) > Cr (0.86). Cr belongs to the precaution level; Tl belongs to the light pollution level. Specifically, the distribution of sampling sites with P_n values higher than 0.7 were N10, N13, N25, and N26 for Cr, N21 and N31 for Tl, respectively. The results indicate the presence of Cr and Tl contamination in the Nianchu River Basin, which could be associated with the region's agriculture and geological composition [27,28]. Compared with the evaluation results of the water quality pollution status for each monitoring region of the YZRB using P_i, P_n comprehensively considered the more seriously polluted indicators, which reflected the degree of water pollution, and also considered the average value of P_i, supplementing the shortcomings of P_i. Therefore, P_n is more comprehensive than P_i for water quality evaluation.

Table 2. Comparison of average concentrations of trace elements in the YZRB (unit: μg/L).

Trace Elements	Fe	V	Be	Ti	Mo	Se	Cd	Zn	Cu	Ni	Co	Mn	Cr	Ba	Tl	Pb	Hg	As	Sb	References
Min	4.5	<5	0.03	0.4	0.096	<0.25	0.06	0.8	0.14	0.8	0.046	0.06	0.09	1	<0.01	0.007	<0.4	0.12	0.078	This study
Max	700	<5	1.3	150	4.2	0.29	0.22	29	28	9.9	1.35	32.9	55	92	0.2	1.1	<0.4	28	3.5	
Mean	101.107	<5	0.143	9.711	1.001	<0.25	0.074	2.507	2.046	2.888	0.249	3.261	4.903	16.851	0.043	0.177	<0.4	2.335	0.639	
Mean	14.925	–	–	0.938	1.518	–	–	–	–	–	–	16.829	–	–	–	–	–	–	–	[21]
Mean	16.628	–	–	0.621	1.673	–	–	–	–	–	–	4.12	–	–	–	–	–	24.48	–	[22]
Mean	–	–	–	–	1.4	–	3.6	20.5	1.9	2.8	–	30.7	2.8	15.3	–	15.8	–	–	4.3	[3]
Mean	–	–	–	7.8	1.2	–	1	9.8	1.4	–	–	12.8	2.7	12	<0.007	5.6	3.225	10.5	3.4	[23]
Mean	20.592	–	–	0.368	0.821	–	0.088	10.8	1.124	–	–	54.86	0.128	43.285	0.014	0.557	0.023	0.862	–	[24]
Mean	30	–	–	7.51	1.83	0.05	0.04	0.53	0.76	2.4	0.37	5.43	4.01	–	–	0.13	0.03	12.2	–	[25]
World Mean	66	–	0.009	0.489	0.42	0.07	0.08	0.6	1.48	0.801	0.148	34	0.7	23	–	0.079	–	0.62	0.07	[26]

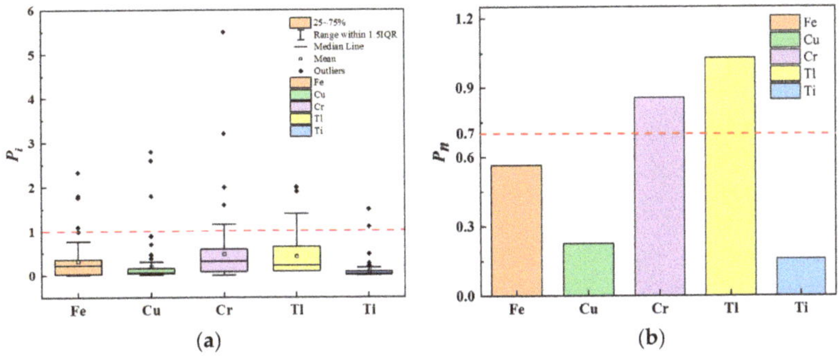

Figure 2. Pollution index of trace elements in the YZRB: (**a**) Single-factor index (P_i); (**b**) multi-factor-integrated pollution index (P_n).

3.3. Spatial Distribution

To present the spatial distribution characteristics of each trace element more clearly, the geographic information system (GIS) method [29,30] was used to perform spatial interpolation analysis on the 16 trace elements, and the spatial distribution graph is shown in Figure 3. Se, Hg, and V were not shown due to weak variability. As, Ba, Mo, and Sb were mainly distributed in Duilongqu (right tributary of the Lhasa River); Zn and Cu were mainly distributed in the Lhasa River; Fe, Mn, Ti, Co, Cr, Tl, and Be were mainly distributed in the Nianchu River Basin. The spatial distribution of Zn and Cu was consistent with the distribution of mines. The distribution of mining areas was mainly concentrated in Duilongdeqing County and Mozhugongka County in Lhasa City [31]. The spatial distribution of As was consistent with the distribution of geothermal resources in the Lhasa River Basin [32,33]. Due to the input of geothermal water, the concentration of As in the Lhasa River Basin was high [34,35]. The trace elements of Mn, Fe, Ti, Pb, and Co were speculated to be weathering of soils input around the river [21]. Cr and Ti were mainly distributed in the Nianchu River Basin, possibly owing to agricultural activities [36,37]. Spatial distribution provides more intuitive assistance when analyzing the source of elements.

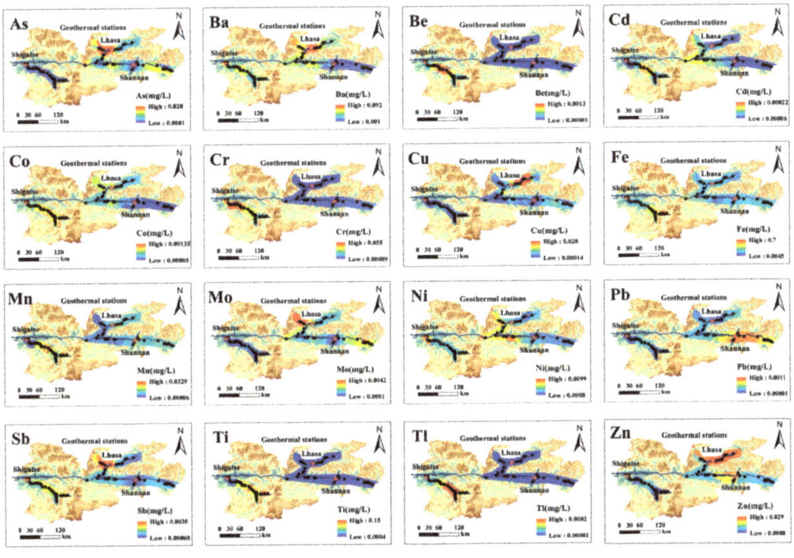

Figure 3. Spatial distribution of trace elements.

3.4. Source Apportionment

3.4.1. Correlation Analysis

Analyzing correlations between trace elements can provide a better understanding of their affinity, and a strong affinity between them indicates that these sources are likely to be the same [38,39]. Due to the concentrations of V and Hg being below the detection limits (0.005 and 0.0004 mg/L, correspondingly), the correlation between Hg, V, and other elements was not discussed. As presented in Figure 4a, there were significant correlations among As, Sb, and Mo ($p < 0.01$); Fe, Mn, Ti, Pb, Ni, Co, and Ba ($p < 0.01$), implying their affinity. Pearson correlation analysis can assist PMF in traceability. Although correlation analysis can demonstrate the affinity between elements, it cannot reveal source contributions. Therefore, it is necessary to use the PMF model for source contribution analysis.

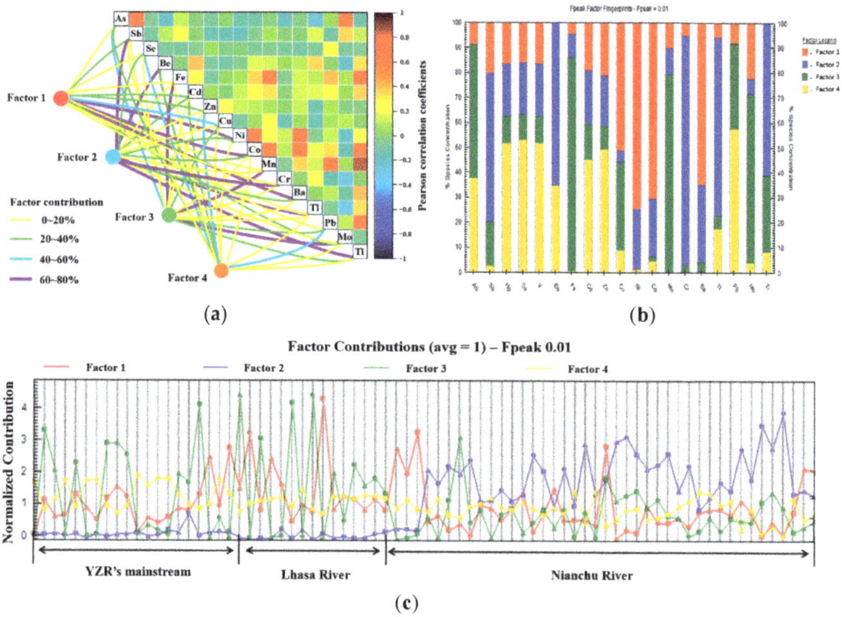

Figure 4. The sources of trace elements were analyzed via PMF: (**a**) Pearson correlation coefficient; (**b**) factor analysis and contribution rate; (**c**) the normalized contributions.

3.4.2. Source Analysis of PMF Model

Through PMF model analysis, four contribution factors were found in the YZRB. Clearly, Figure 4b delineates the contribution concentration and contribution ratio of various trace elements as factors. Generally, a high contribution concentration that simultaneously meets a high contribution ratio can be identified as the dominant contribution factor [40,41].

Factor 1 (F1) was mainly loaded on Cu, Ni, Co, Zn, and Ba. Based on spatial distribution, there are copper mines distributed upstream of the Lhasa River [42,43]; therefore, the source of Cu and Zn is the mining wastewater discharge. Also, Ni, Co, and Ba are released into the aquatic environment as a result of the resuspension and dissolution of river sediments [44–46]. Therefore, F1 can be attributed to the source of mining wastewater and sediments (16.283%). Factor 2 was dominated by Cr, Sb, and Ti. Previous studies have shown that the concentration of Cr in fertilizers and pesticides is generally high (11.5–94.6 mg/kg) [47]. Based on the previous spatial distribution results, Cr had high concentrations in the Shigatse greenhouse area. In addition, certain agricultural activities also use titanium-containing compounds [48,49]. Therefore, F2 can be attributed to the source of agricultural activities (11.436%). Factor 3 was dominated by Fe, As, Mn, and Mo. Interestingly, both As and Mo elements exhibited a high value of distribution in the Duilongqu Basin, where geothermal resources are widely distributed. Considering that YZRB is an ancient karst region, with huge potential for Fe and Mn deposits [16,50–52], F3 could be identified as a lithological source and geothermal wastewater (47.418%). Factor 4 (F4) was mainly loaded on Hg, Se, V, and Zn. As mentioned above, the spatial distribution for Se, Hg, and V showed weak variability, with detection concentrations appearing relatively low. Thus, F4 was predominantly characterized by Zn. Zn mainly exhibits a high-value distribution in the Lhasa River Basin, and this overall regional variation was not significant when manifested by the weathering of soils [53]. Therefore, F4 was primarily from soil-forming rocks (6.374%).

Figure 4c presents the normalized contribution source analysis in each watershed. The results show that the trace elements found in the YZRB were predominantly influenced by the surrounding rock composition (F3 and F4). Meanwhile, the Lhasa River water environment marked both the discharge of mining/geothermal wastewater (F1 and F3). Additionally, agricultural activities (F2) were the chief contributors to trace elements in the Nianchu River.

3.4.3. Cluster Analysis

To visualize the origins of trace elements at each sampling point, cluster analysis was conducted [54]. The diagrams of clustering analysis yielded for the sampling sites are shown in Figure 5.

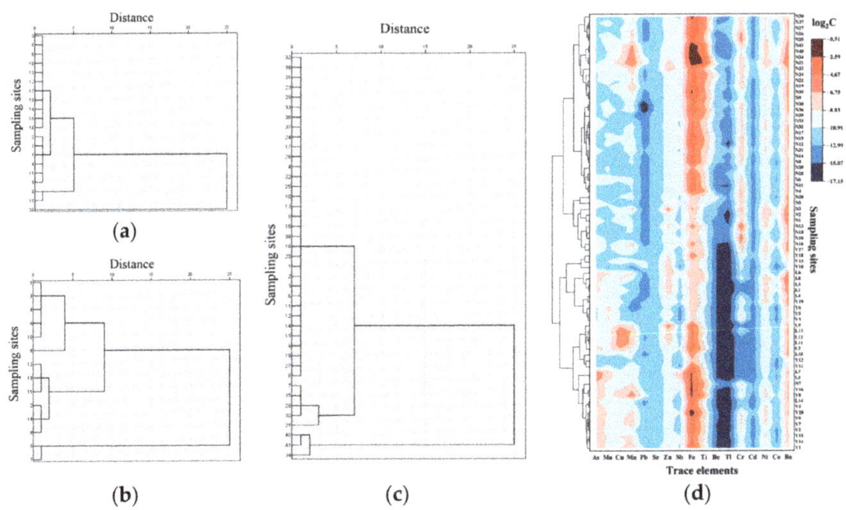

Figure 5. Cluster analysis of sampling sites in the YZRB: (**a**) YZR's mainstream; (**b**) Lhasa River; (**c**) Nianchu River; (**d**) clustering heat map of the whole sampling sites in the YZRB. C represents the concentration of trace elements (unit: mg/L).

In Figure 5a, the sampling sites in the YZR's mainstream can be divided into three categories. There are seventeen sampling sites (Y1–7, Y7–15, and Y17–19), two sampling sites (Y8 and Y16), and one sampling site (Y20) in these three categories, respectively. The sampling sites with the highest proportion were all least affected by human activities. The sampling sites of Y8 and Y16 were located near the highway and were greatly affected by traffic sources. At sampling point Y20, a substantial sediment deposit accumulated on the riverbed, leading to the re-release of Fe from the sediment. Therefore, the main source of the YZR's mainstream is the surrounding rock composition.

In Figure 5b, the sampling sites in the Lhasa River can be divided into three categories. There are six sampling sites (L1, L3, L4, L6, L8, and L10) in the first category. These sampling sites are located downstream of Lhasa city and are affected by the input of production and domestic sewage. There are six sampling points (L2, L9, and L11–14) in the second category. Significantly, sampling sites from L11 to L14 were located downstream of the mining area (Mozhugongka County). Mining wastewater input was the most important source. The third type includes two sampling sites (L5 and L7), both of which were downstream of Duilongqu, where geothermal resources were widely distributed. Therefore, geothermal wastewater from Yangbajing produced a significant effect [55]. In conclusion, the main source of the Lhasa River was the discharge of mining and geothermal wastewater.

In Figure 5c, the sampling sites of Nianchu River are divided into three categories. There are 33 sampling sites in the first category, N1–6, N8–20, N22–25, N27–29, N31–33, and N36–39. Most of these sites are surrounded by greenhouses. Based on the pollution assessment results, the P_i values of Cr at these sites are significantly high. Thus, the main sources of pollution at these sampling sites are agricultural activities.

In Figure 5d, a clustering heat map of sampling points in the whole YZRB is shown. It can be seen from the color that the concentration of Fe is higher than that of other trace elements in the whole basin, which is consistent with the factor distribution and contribution rate plot of PMF.

4. Conclusions

In conclusion, this study investigated the concentration levels, spatial distribution, and sources of 19 trace elements in the YZRB. Through the evaluation of the Nemerow pollution index, the YZRB was considered to be in an unpolluted state with limited human impact. Due to their high P_i values, five trace elements (Cr, Cu, Fe, Tl, and Ti) need to be continuously studied in future research. The spatial distribution of Cu and Zn is consistent with the distribution of mines, while the spatial distribution of As is consistent with the distribution of geothermal resources. Cr is mainly distributed in the Nianchu River Basin, and Fe is uniformly distributed throughout the basin. The use of PMF to analyze trace element sources, conduct correlation analysis, and perform cluster analysis is an innovative way to assist with its traceability. Cluster analysis combined with PMF source analysis showed that the trace elements found in the YZRB were predominantly influenced by the lithological composition of host rocks, with the Lhasa River showing the discharge of mining and geothermal wastewater and the Nianchu River being affected by agricultural activities. Furthermore, the entire basin was subjected to the influence of soil-forming rocks. This study comprehensively examines trace element spatial characteristics in the YZRB, utilizing surface, line, and point perspectives. It employs Kriging for surface analysis to study trace element distribution. Shifting to a line approach, it conducts a comparative analysis of contributions from four sources in three river sections, using normalized contribution for precise source tracing. Finally, at the point level, cluster analysis is employed to understand factors behind trace element concentration outliers at various sampling points. This study seamlessly transitions from a surface to line and point perspective and is expected to provide a foundational reference for further research on the spatial distribution and sources of trace elements in rivers on the QTP.

Author Contributions: Conceptualization, F.X. and Y.Z.; methodology, F.X.; software, F.X.; validation, F.X., Y.Z. and Q.Z.; formal analysis, F.X.; investigation, Y.Z.; resources, D.B.; data curation, F.X.; writing—original draft preparation, F.X.; writing—review and editing, F.X. and Q.Z.; visualization, F.X.; supervision, Q.Z.; project administration, D.B.; funding acquisition, D.B. All authors have read and agreed to the published version of the manuscript.

Funding: This research was funded by the Second Comprehensive Scientific Investigation Project of Qinghai-Tibet Plateau (No.2019QZKK0603), the National Natural Science Foundation of China (No.22266032, No.41967059), State Key Laboratory of Environmental Chemistry and Ecotoxicology, Research Center for Eco-Environmental Sciences, Chinese Academy of Sciences (No.KF2021–05), Central Financial Support Special Funds for Local Colleges and Universities ([2023] No.1).

Data Availability Statement: The datasets used and analyzed during the current study are available from the corresponding author on reasonable request.

Conflicts of Interest: The authors declare no conflict of interest.

References

1. Immerzeel, W.W.; Lutz, A.F.; Andrade, M.; Bahl, A.; Biemans, H.; Bolch, T.; Hyde, S.; Brumby, S.; Davies, B.J.; Elmore, A.C.; et al. Importance and vulnerability of the world's water towers. *Nature* **2020**, *577*, 364–369. [CrossRef]
2. Li, Y.; Han, F.; Zhou, H. Assessment of terrestrial ecosystem sensitivity and vulnerability in Tibet. *J. Resour. Ecol.* **2017**, *8*, 526–537.
3. Qu, B.; Zhang, Y.; Kang, S.; Sillanpää, M. Water chemistry of the southern Tibetan Plateau: An assessment of the Yarlung Tsangpo river basin. *Environ. Earth Sci.* **2017**, *76*, 74. [CrossRef]
4. Wu, J.; Duan, D.; Lu, J.; Luo, Y.; Wen, X.; Guo, X.; Boman, B.J. Inorganic pollution around the Qinghai-Tibet Plateau: An overview of the current observations. *Sci. Total Environ.* **2016**, *550*, 628–636. [CrossRef]
5. Dutta, S.; Dutta, P.; Devi, U.; Sarmaa, K.P. Distribution and pollution status of metals in some water bodies of mid-Brahmaputra valley. *Desalin. Water Treat.* **2017**, *65*, 215–223. [CrossRef]
6. Wang, C.; Zhou, H.; Kuang, X.; Hao, Y.; Shan, J.; Chen, J.; Li, L.; Feng, Y.; Zou, Y.; Zheng, Y. Water quality and health risk assessment of the water bodies in the Yamdrok-tso basin, southern Tibetan Plateau. *J. Environ. Manag.* **2021**, *300*, 113740. [CrossRef]
7. Zheng, T.; Deng, Y.; Lin, H.; Xie, Y.; Pei, X. Hydrogeochemical controls on As and B enrichment in the aqueous environment from the Western Tibetan Plateau: A case study from the Singe Tsangpo River Basin. *Sci. Total Environ.* **2022**, *817*, 152978. [CrossRef]
8. Huang, X.; Sillanpää, M.; Gjessing, E.T.; Peräniemi, S.; Vogt, R.D. Environmental impact of mining activities on the surface water quality in Tibet: Gyama valley. *Sci. Total Environ.* **2010**, *408*, 4177–4184. [CrossRef]
9. Kara, G.T.; Kara, M.; Bayram, A.; Gündüz, O. Assessment of seasonal and spatial variations of physicochemical parameters and trace elements along a heavily polluted effluent-dominated stream. *Environ. Monit. Assess.* **2017**, *189*, 585. [CrossRef]
10. Zhan, S.; Wu, J.; Jin, M. Hydrochemical characteristics, trace element sources, and health risk assessment of surface waters in the Amu Darya Basin of Uzbekistan, arid Central Asia. *Environ. Sci. Pollut. Res.* **2022**, *29*, 5269–5281. [CrossRef]
11. Cong, Z.; Kang, S.; Zhang, Y.; Li, X. Atmospheric wet deposition of trace elements to central Tibetan Plateau. *Appl. Geochem.* **2010**, *25*, 1415–1421. [CrossRef]
12. Jin, Z.; You, C.; Yu, T.; Wang, B. Elemental distribution in the topsoil of the Lake Qinghai catchment, NE Tibetan Plateau, and the implications for weathering in semi-arid areas. *J. Geochem. Explor.* **2015**, *152*, 1–9.
13. State Environmental Protection Administration of the People's Republic of China. *Environmental Quality Standards for Surface Water*; State Environmental Protection Administration of the People's Republic of China: Beijing, China, 2002. Available online: https://www.mee.gov.cn/ywgz/fgbz/bz/bzwb/shjbh/shjzlbz/200206/t20020601_66497.shtml (accessed on 26 May 2023).
14. Xia, Z.; Zhang, J.; Yan, Y.; Zhang, W.; Zhao, Z. Heavy metals in suspended particulate matter in the Yarlung Tsangpo River, Southwest China. *GeoSyst. Geoenv.* **2022**, 100160. [CrossRef]
15. Shi, D.; Tan, H.; Chen, X.; Rao, W.; Issombo, H.E.; Basang, R. Temporal and spatial variations of runoff composition revealed by isotopic signals in Nianchu River catchment, Tibet. *J. Hydrol. Environ. Res.* **2021**, *37*, 1–12. [CrossRef]
16. Zhou, X.; Ao, Y.; Jiang, X.; Yang, S.; Hu, Y.; Wang, X.; Zhang, J. Water use efficiency of China's karst ecosystems: The effect of different ecohydrological and climatic factors. *Sci. Total Environ.* **2023**, *905*, 167069. [CrossRef]
17. Su, K.; Wang, Q.; Li, L.; Cao, R.; Xi, Y. Water quality assessment of Lugu Lake based on Nemerow pollution index method. *Sci. Rep.* **2022**, *12*, 13613. [CrossRef]
18. Belle, G.; Schoeman, Y.; Oberholster, P. Potential toxic-element pollution in surface water and its implications for aquatic and human health: Source-pathway-receptor model. *Water* **2023**, *15*, 3100. [CrossRef]
19. Paatero, P.; Tapper, U. Positive matrix factorization: A non-negative factor model with optimal utilization of error estimates of data values. *Environmetrics* **2010**, *5*, 111–126. [CrossRef]
20. Wetzel, R.G. Lake and river ecosystems. In *Limnology*; Academic Press: San Diego, CA, USA, 2002.
21. Huang, X.; Sillanpää, M.; Gjessing, E.T.; Vogt, R.D. Water quality in the Tibetan Plateau: Major ions and trace elements in the headwaters of four major Asian rivers. *Sci. Total Environ.* **2009**, *407*, 6242–6254. [CrossRef]
22. Huang, X.; Sillanpää, M.; Gjessing, E.T.; Peräniemi, S.; Vogt, R.D. Water quality in the southern Tibetan Plateau: Chemical evaluation of the Yarlung Tsangpo (Brahmaputra). *River Res. Appl.* **2011**, *27*, 113–121. [CrossRef]
23. Qu, B.; Zhang, Y.; Kang, S.; Sillanpää, M. Water quality in the Tibetan Plateau: Major ions and trace elements in rivers of the "Water Tower of Asia". *Sci. Total Environ.* **2019**, *649*, 571–581. [CrossRef]
24. Wu, J.; Zhao, Z.; Meng, j.; Zhou, T. Analysis on water quality of the Tibetan Plateau based on the major ions and trace elements in the Niyang River Basin. *Appl. Ecol. Environ. Res.* **2020**, *18*, 3729–3740. [CrossRef]
25. Liu, J.; Guo, H. Hydrochemical characteristics and ion source analysis of the Yarlung Tsangpo River Basin. *Water* **2023**, *15*, 537. [CrossRef]
26. Gaillardet, J.; Viers, J.; Dupré, B. Trace elements in river waters. *Treatise Geochem.* **2003**, *5*, 225–272.
27. Du, M.; Zhang, Q.; Ren, P.; Gao, S.; Bu, D. Distribution of soil heavy metals and ecological risk assessment of agricultural land in the Nianchu River Basin, Tibet. *J. Environ. Eng. Technol.* **2022**, *12*, 1618–1625.

28. Guo, J.; Cao, Y.; Luo, Z.; Fang, H.; Chen, Z.; Wang, D.; Xu, F.; Yan, C. Distribution, fractions, and potential release of thallium in acidic soils nearby a waste copper mining site from southern China. *Environ. Sci. Pollut. Res.* **2018**, *25*, 17980–17988. [CrossRef]
29. Madadzada, A.I.; Badawy, W.M.; Hajiyeva, S.R.; Veliyeva, Z.T.; Hajiyev, O.B.; Shvetsova, M.S.; Frontasyeva, M.V. Assessment of atmospheric deposition of major and trace elements using neutron activation analysis and GIS technology: Baku-Azerbaijan. *Microchem. J.* **2017**, *147*, 605–614. [CrossRef]
30. Li, F.; Qiu, Z.; Zhang, J.; Liu, C.; Cai, Y.; Xiao, M. Spatial distribution and fuzzy health risk assessment of trace elements in surface water from Honghu Lake. *Int. J. Environ. Res. Public Health* **2017**, *14*, 1011. [CrossRef]
31. Bu, D.; Xu, Z.; Wu, J.; Li, M.; Dan, Z.; De, J. Study on the impact of mineral processing plant on environment in Lhasa river catchments. *J. Tibet Univ.* **2009**, *24*, 33–38.
32. Zhang, Y.F.; Tan, H.B.; Zhang, W.J.; Huang, J.Z.; Zhang, Q. A new geochemical perspective on hydrochemical evolution of the Tibetan geothermal system. *Geochem. Int.* **2015**, *53*, 1090–1106. [CrossRef]
33. Li, C.; Kang, S.; Chen, P.; Zhang, Q.; Mi, J.; Gao, S.; Sillanpää, M. Geothermal spring causes arsenic contamination in river waters of the southern Tibetan Plateau, China. *Environ. Earth Sci.* **2014**, *71*, 4143–4148. [CrossRef]
34. Guo, Q.; He, T.; Wu, Q.; Liu, M. Constraints of major ions and arsenic on the geological genesis of geothermal water: Insight from a comparison between Xiong'an and Yangbajain, two hydrothermal systems in China. *Appl. Geochem.* **2020**, *117*, 104589. [CrossRef]
35. Zhang, J.; Yan, Y.; Zhao, Z.; Li, X.; Guo, J.; Ding, H.; Cui, L.; Meng, J.; Liu, C. Spatial and seasonal variations of dissolved arsenic in the Yarlung Tsangpo River, southern Tibetan Plateau. *Sci. Total Environ.* **2021**, *760*, 143416. [CrossRef] [PubMed]
36. Xing, W.; Wei, L.; Ma, W.; Li, J.; Liu, X.; Hu, J.; Wang, X. Geochemistry and sources apportionment of major ions and dissolved heavy metals in a small watershed on the Tibetan Plateau. *Water* **2022**, *14*, 3856. [CrossRef]
37. Nguyen, N.T.T.; Nguyen, L.M.; Nguyen, T.T.T.; Liew, R.K.; Nguyen, D.T.C.; Tran, T.V. Recent advances on botanical biosynthesis of nanoparticles for catalytic, water treatment and agricultural applications: A review. *Sci. Total Environ.* **2022**, *827*, 154160. [CrossRef] [PubMed]
38. Li, S.; Zhang, Q. Spatial characterization of dissolved trace elements and heavy metals in the upper Han River (China) using multivariate statistical techniques. *J. Hazard. Mater.* **2010**, *176*, 579–588. [CrossRef] [PubMed]
39. Leonardo, R.D.; Adelfio, G.; Bellanca, A.; Chiodi, M.; Mazzola, S. Analysis and assessment of trace element contamination in offshore sediments of the Augusta Bay (SE Sicily): A multivariate statistical approach based on canonical correlation analysis and mixture density estimation approach. *J. Sea Res.* **2014**, *85*, 428–442. [CrossRef]
40. Pekey, H.; Doğan, G. Application of positive matrix factorisation for the source apportionment of heavy metals in sediments: A comparison with a previous factor analysis study. *Microchem. J.* **2013**, *106*, 233–237. [CrossRef]
41. Comero, S.; Vaccaro, S.; Locoro, G.; Capitani, L.D.; Gawlik, B.M. Characterization of the Danube River sediments using the PMF multivariate approach. *Chemosphere* **2014**, *95*, 329–335. [CrossRef] [PubMed]
42. Zeng, Y.; Chen, J.; Xu, J.; Lei, M.; Xiong, Q. Origin of Miocene Cu-bearing porphyries in the Zhunuo region of the southern Lhasa subterrane: Constraints from geochronology and geochemistry. *Gondwana Res.* **2017**, *41*, 51–64. [CrossRef]
43. Zheng, W.; Tang, J.; Zhong, K.; Ying, L.; Leng, Q.; Ding, S.; Lin, B. Geology of the Jiama porphyry copper-polymetallic system, Lhasa Region, China. *Ore Geol. Rev.* **2016**, *74*, 151–169. [CrossRef]
44. Wang, K.; Zhang, L.; Jiang, X.; Zhao, L.; Wang, S. Determination of background value and potential ecological risk of heavy metals in sediments of a deep Plateau Lake. *Res. Environ. Sci.* **2018**, *31*, 2124–2132.
45. Bai, J.; Li, C.; Kang, S.; Chen, P.; Wang, J. Chemical speciation and risk assessment of heavy metals in the middle part of Yarlung Zangbo surface sediments. *Environ. Sci.* **2014**, *5*, 3346–3351.
46. Dupré, B.; Gaillardet, J.; Rousseau, D.; Allègre, C.J. Major and trace elements of river-borne material: The Congo Basin. *Geochim. Cosmochim. Acta* **1996**, *60*, 1301–1321. [CrossRef]
47. Su, J. The Conventional Pesticide Products Cr, Pb, Cd, As, Hg Heavy Metal Analysis in Gan Su Province. Master's Thesis, Lanzhou University, Lanzhou, China, 2018.
48. Vidmar, J.; Zuliani, T.; Milačič, R.; Ščančar, J. Following the occurrence and origin of titanium dioxide nanoparticles in the Sava River by single particle ICP-MS. *Water* **2022**, *14*, 959. [CrossRef]
49. Rodríguez-González, V.; Terashima, C.; Fujishima, A. Applications of photocatalytic titanium dioxide-based nanomaterials in sustainable agriculture. *J. Photochem. Photobiol. C Photochem. Rev.* **2019**, *40*, 49–67. [CrossRef]
50. Ma, J.; Li, X.; Zhang, C.; Fu, C.; Xie, X.; Wang, X.; Li, X.; Zhang, D.; Bai, Z.; Wang, Z. Recharge source, mode and development potential of typical tectonic karst groundwater in the eastern Qinghai-Xizang Plateau. *Geol. China* **2022**, 1–20. Available online: http://kns.cnki.net/kcms/detail/11.1167.P.20220822.1417.016.html (accessed on 1 May 2023).
51. Pu, T.; Kong, Y.; Kang, S.; Shi, X.; Zhang, G.; Wang, S.; Cao, B.; Wang, K.; Hua, H.; Chen, P. New insights into trace elements in the water cycle of a karst-dominated glacierized region, southeast Tibetan Plateau. *Sci. Total Environ.* **2021**, *751*, 141725. [CrossRef]
52. Wen, Y.; Li, W.; Yang, Z.; Zhang, Q.; Ji, J. Enrichment and source identification of Cd and other heavy metals in soils with high geochemical background in the karst region, Southwestern China. *Chemosphere* **2020**, *245*, 125620. [CrossRef] [PubMed]
53. Pan, Y.; Chen, M.; Wang, X.; Chen, Y.; Dong, K. Ecological risk assessment and source analysis of heavy metals in the soils of a lead-zinc mining watershed area. *Water* **2023**, *15*, 113. [CrossRef]

54. Hussain, M.; Ahmed, S.M.; Abderrahman, W. Cluster analysis and quality assessment of logged water at an irrigation project, eastern Saudi Arabia. *J. Environ. Manag.* **2008**, *86*, 297–307. [CrossRef] [PubMed]
55. Wang, Y.; Li, L.; Wen, H.; Hao, Y. Geochemical evidence for the nonexistence of supercritical geothermal fluids at the Yangbajing geothermal field, southern Tibet. *J. Hydrol.* **2022**, *604*, 127243. [CrossRef]

MDPI AG
Grosspeteranlage 5
4052 Basel
Switzerland
Tel.: +41 61 683 77 34

MDPI Books Editorial Office
E-mail: books@mdpi.com
www.mdpi.com/books

www.ingramcontent.com/pod-product-compliance
Lightning Source LLC
Chambersburg PA
CBHW062050220926
51850CB00010B/2720

* 9 7 8 3 7 2 5 8 7 6 0 1 3 *